Assessment of Sea-Turtle Status and Trends

Integrating Demography and Abundance

Committee on the Review of Sea-Turtle Population Assessment Methods

Ocean Studies Board

Division on Earth and Life Studies

NATIONAL RESEARCH COUNCIL
OF THE NATIONAL ACADEMIES

THE NATIONAL ACADEMIES PRESS
Washington, D.C.
www.nap.edu

THE NATIONAL ACADEMIES PRESS 500 Fifth Street, N.W. Washington, DC 20001

NOTICE: The project that is the subject of this report was approved by the Governing Board of the National Research Council, whose members are drawn from the councils of the National Academy of Sciences, the National Academy of Engineering, and the Institute of Medicine. The members of the committee responsible for the report were chosen for their special competences and with regard for appropriate balance.

This study was supported by Contract No. DG133R08CQ0062 between the National Academy of Sciences and the National Oceanic and Atmospheric Administration. Any opinions, findings, conclusions, or recommendations expressed in this publication are those of the authors and do not necessarily reflect the views of the organizations or agencies that provided support for the project.

International Standard Book Number-13: 978-0-309-15255-6
International Standard Book Number-10: 0-309-15255-0

Cover: The front cover images include five of the six species of sea turtles found in U.S. waters (from top to bottom and left to right): leatherback (provided by Guillaume Feuillet, Association Kwata), Kemp's ridley (provided by Selina Heppell, Oregon State University), hawksbill (provided by the National Oceanic and Atmospheric Administration), olive ridley (provided by Guillaume Feuillet, Association Kwata), and green (provided by Claire Fackler, National Oceanic and Atmospheric Administration). The back cover image is the sixth species, which is the loggerhead (provided by William Precht, National Oceanic and Atmospheric Administration).

Additional copies of this report are available from the National Academies Press, 500 Fifth Street, N.W., Lockbox 285, Washington, DC 20055; (800) 624-6242 or (202) 334-3313 (in the Washington metropolitan area); Internet, http://www.nap.edu.

Printed in the United States of America

THE NATIONAL ACADEMIES

Advisers to the Nation on Science, Engineering, and Medicine

The **National Academy of Sciences** is a private, nonprofit, self-perpetuating society of distinguished scholars engaged in scientific and engineering research, dedicated to the furtherance of science and technology and to their use for the general welfare. Upon the authority of the charter granted to it by the Congress in 1863, the Academy has a mandate that requires it to advise the federal government on scientific and technical matters. Dr. Ralph J. Cicerone is president of the National Academy of Sciences.

The **National Academy of Engineering** was established in 1964, under the charter of the National Academy of Sciences, as a parallel organization of outstanding engineers. It is autonomous in its administration and in the selection of its members, sharing with the National Academy of Sciences the responsibility for advising the federal government. The National Academy of Engineering also sponsors engineering programs aimed at meeting national needs, encourages education and research, and recognizes the superior achievements of engineers. Dr. Charles M. Vest is president of the National Academy of Engineering.

The **Institute of Medicine** was established in 1970 by the National Academy of Sciences to secure the services of eminent members of appropriate professions in the examination of policy matters pertaining to the health of the public. The Institute acts under the responsibility given to the National Academy of Sciences by its congressional charter to be an adviser to the federal government and, upon its own initiative, to identify issues of medical care, research, and education. Dr. Harvey V. Fineberg is president of the Institute of Medicine.

The **National Research Council** was organized by the National Academy of Sciences in 1916 to associate the broad community of science and technology with the Academy's purposes of furthering knowledge and advising the federal government. Functioning in accordance with general policies determined by the Academy, the Council has become the principal operating agency of both the National Academy of Sciences and the National Academy of Engineering in providing services to the government, the public, and the scientific and engineering communities. The Council is administered jointly by both Academies and the Institute of Medicine. Dr. Ralph J. Cicerone and Dr. Charles M. Vest are chair and vice chair, respectively, of the National Research Council.

www.national-academies.org

COMMITTEE ON THE REVIEW OF SEA-TURTLE POPULATION ASSESSMENT METHODS

KAREN A. BJORNDAL (*Chair*), University of Florida, Gainesville
BRIAN W. BOWEN, University of Hawaii
MILANI CHALOUPKA, Ecological Modelling Services Pty Ltd, St. Lucia, Australia
LARRY B. CROWDER, Duke University, Beaufort, North Carolina
SELINA S. HEPPELL, Oregon State University, Corvallis
CYNTHIA M. JONES, Old Dominion University, Norfolk, Virginia
MOLLY E. LUTCAVAGE, University of Massachusetts, Amherst
ANDREW R. SOLOW, Woods Hole Oceanographic Institution, Massachusetts
BLAIR E. WITHERINGTON, Florida Fish and Wildlife Conservation Commission, Melbourne Beach

Staff

JODI BOSTROM, Associate Program Officer
SUSAN PARK, Senior Program Officer (until December 31, 2009)
DAVID POLICANSKY, Scholar
JEREMY JUSTICE, Senior Program Assistant
NORMAN GROSSBLATT, Senior Editor

SHUBHA BANSKOTA, Financial Associate
PAMELA LEWIS, Administrative Coordinator
SHERRIE FORREST, Research Associate
HEATHER CHIARELLO, Senior Program Assistant
JEREMY JUSTICE, Senior Program Assistant

Acknowledgments

This report was greatly enhanced by the participants of the meetings held as part of the committee's study. The committee acknowledges the efforts of those who gave presentations at meetings: Steve Murawski (National Oceanic and Atmospheric Administration), Barbara Schroeder (National Oceanic and Atmospheric Administration), Sandy MacPherson (U.S. Fish and Wildlife Service), Earl Possardt (U.S. Fish and Wildlife Service), Yan Jiao (Virginia Polytechnic Institute and State University), Mark Maunder (Inter-American Tropical Tuna Commission), Paul Wade (National Oceanic and Atmospheric Administration), and Bryan Wallace (Conservation International). Their talks helped to set the stage for fruitful discussions in the closed sessions that followed.

This report has been reviewed in draft form by individuals chosen for their diverse perspectives and technical expertise, in accordance with procedures approved by the National Research Council's Report Review Committee. The purpose of this independent review is to provide candid and critical comments that will assist the institution in making its published report as sound as possible and to ensure that the report meets institutional standards for objectivity, evidence, and responsiveness to the study charge. The review comments and draft manuscript remain confidential to protect the integrity of the deliberative process. We wish to thank the following individuals for their review of this report:

RICHARD T. DURRETT, Cornell University, Ithaca, New York
MASAMI FUJIWARA, Texas A&M University, College Station

MATTHEW GODFREY, Sea Turtle Project, Beaufort, North Carolina
BRENDAN GODLEY, University of Exeter, Tremough Campus, United Kingdom
MARTÍN HALL, Inter-American Tropical Tuna Commission, La Jolla, California
WILLIAM L. KENDALL, U.S. Geological Survey, Laurel, Maryland
JACK MUSICK, College of William and Mary, Gloucester, Virginia
EUGENIA NARO-MACIEL, American Museum of Natural History, New York
MICHAEL ORBACH, Duke University, Beaufort, North Carolina
JOSEPH M. QUATTRO, University of South Carolina, Columbia
H. BRADLEY SHAFFER, University of California, Davis
MELISSA SNOVER, National Park Service, Denali Park, Alaska
PATRICK SULLIVAN, Cornell University, Ithaca, New York
JEANETTE WYNEKEN, Florida Atlantic University, Boca Raton

Although the reviewers listed above have provided many constructive comments and suggestions, they were not asked to endorse the conclusions or recommendations nor did they see the final draft of the report before its release. The review of this report was overseen by **John J. Magnuson** (University of Wisconsin) and **Robert T. Paine** (University of Washington). Appointed by the National Research Council, they were responsible for making certain that an independent examination of this report was carried out in accordance with institutional procedures and that all review comments were carefully considered. Responsibility for the final content of this report rests entirely with the authoring committee and the institution.

Contents

Summary

INTRODUCTION AND BACKGROUND

Long lifespans and wide-ranging migrations make the seven species of sea turtles difficult to monitor. They are susceptible to many sources of mortality, including direct and incidental "takes" (basically any potential effect on a turtle or its behavior [50 CFR 17.3]) resulting from coastal and oceanic human activities. All six of the species that occur in U.S. waters[1]—loggerhead (*Caretta caretta*), green (*Chelonia mydas*), hawksbill (*Eretmochelys imbricate*), Kemp's ridley (*Lepidochelys kempii*), olive ridley (*Lepidochelys olivacea*), and leatherback (*Dermochelys coriacea*)—are listed as endangered or threatened under the Endangered Species Act so their direct harvest is prohibited although incidental take is permitted under some circumstances. (The seventh sea-turtle species is the flatback [*Natator depressus*], which is found only in the waters around Australia, Papua New Guinea, and Indonesia.) Accurate assessments are necessary to evaluate the status and trends of populations and the effects of incidental takes and to assess the value of implemented protections of specific populations.

Sea turtles migrate across whole ocean basins; therefore, population assessments require an international context. Activities throughout the world's oceans—including development on nesting beaches, killing of turtles for food, and incidental capture in commercial and subsistence

[1] U.S. waters not only refers to waters around U.S. states but also waters around U.S. territories, such as American Samoa, Puerto Rico, Northern Mariana Island, Guam, the U.S. Virgin Islands, and Palmyra Atoll.

fisheries—contribute to sea-turtle declines and affect populations in U.S. waters. Data needed for accurate assessments of most populations are not available, and this prohibits diagnostic evaluation that can benefit management. In light of that problem, the National Marine Fisheries Service (NMFS) of the National Oceanic and Atmospheric Administration asked the National Research Council's (NRC) Ocean Studies Board for advice on methods for improving sea-turtle population assessments. See Box S.1 for the committee's full statement of task.

In response, the NRC appointed a committee of experts. The committee held two public meetings during which it received briefings from NMFS, the U.S. Fish and Wildlife Service (USFWS), and a number of other experts in sea-turtle biology and population assessments. In addition, the committee reviewed the available literature, met in closed sessions, and participated in several conference calls to complete work on its report. The report is intended to help NMFS and USFWS to improve population assessments of sea turtles; NMFS is responsible for the management of sea turtles in the water, and USFWS is responsible for sea turtles on land.

The committee was asked to evaluate current and emerging population assessment techniques that are being applied to provide advice to managers of sea turtles in the United States. Unlike the charge to the committee that prepared the 1990 NRC report (*Decline of the Sea Turtles: Causes and Prevention*), the charge to the present committee was not to review the wide array of threats and management actions related to sea turtles in the United States but rather to focus on the steps necessary to improve the assessments required for federal sea-turtle monitoring and management. This report describes a variety of assessment types and techniques, including beach samples, in-water surveys, genetic analyses, demographic[2] analyses, bycatch (incidental take) information, and aerial surveys; reviews assessment methods; identifies information gaps; and suggests improvements for data collection. The fundamental theme underlying this report is that abundance assessment is essential but that abundance information alone is insufficient to understand the causes underlying trends in sea-turtle populations or to predict future trends. In addition to reliable abundance estimates, it is necessary to understand key demographics. To date, sufficiently complete demographic information has not been used in population assessments of sea turtles in the United States, in large part because it has not been available.

The committee believed that it was beyond its charge to discuss major stresses on sea-turtle populations, such as interactions with fisheries, and

[2] Demographic or vital-rate parameters, such as birth and survival rates, indicate the potential for changes in a population.

Box S.1
Statement of Task

This study will review recent assessments on the status and trends of sea-turtle populations that occur in U.S. waters during all or a portion of their life cycle. The study will evaluate the state of the science and research in terms of population assessment capabilities and data required to improve assessments. The study will review the utility of existing research programs that provide information for assessing and managing sea-turtle populations in the context of current recovery plans. The report will include a discussion of current methods used to assess the status of sea-turtle populations and to estimate known mortality. Recommendations will focus on the research, monitoring, and data needed to improve sea-turtle population assessments in the short- and long-term, such as genetic analyses, telemetry, and mark-recapture studies, taking into account the effectiveness, cost, and timeliness of various data collection methods. The committee will also recommend improvements to existing models, highlight limitations in current methods, identify potential new avenues for modeling, and suggest methods for making sea-turtle population data available for incorporation into a wide range of models and meta-analytical studies.

the potential effects of environmental conditions or external stresses; to detail environmental conditions or regime changes; and to assess the costs of its recommendations. Additionally, this report does not review specific assessments comprehensively, except as illustrative examples of methods and data gaps but does provide a summary of methods used. The committee was not asked to conduct its own assessments of sea-turtle populations but was asked to evaluate the methods used to assess sea-turtle status and trends. That critical distinction was confirmed with NMFS by project staff. As a result, the report does not provide information on the status of sea-turtle populations. The committee recognizes the importance of taking an ecosystem approach to managing sea-turtle populations, but its report focuses on population assessments of single species. Before agencies can undertake ecosystem-based approaches to assessments of sea-turtle populations, substantial information at the single-population or single-species level is needed, as described in this report.

On the basis of its review of the methods used in assessments (see Table 1.2), the committee concludes that most of the modeling and analysis that has been done constitutes a valiant effort to compensate for a debilitating lack of data. The assessment methods that have worked in fishery biology are less successful for turtles because the available data generally are not as complete as they are for many commercial fish species. Filling the large gaps in the available data has far greater promise for improving sea-turtle assessments than refinement of analytical methods. The committee therefore decided that the greatest focus needs to be on the data problem both for the committee's report and for the future activities of the agencies. Developing a rigorous process for assessment of sea-turtle populations is a high priority, but assessments can be more profitably reviewed and refined after better data and a transparent framework for scientific review become available.

ASPECTS OF SEA-TURTLE ASSESSMENTS

Units of Assessment

Understanding the units of assessment in sea turtles requires clarity in the definition of nesting populations. Females show affinity for specific nesting sites, and this potentially causes subpopulations to be vulnerable to extinction. Males breed with females that can have various nesting-site affinities and thus provide male-mediated gene flow among the subpopulations. Because male gene contribution may occur on a larger geographic scale than female gene contribution, that scale defines the geographic upper limits of the nesting populations.

The natural history of sea turtles includes several phases that are difficult to observe directly. In particular, the long generation time and the oceanic habitat of juveniles present major obstacles to studies of immature stages. The genetic identification of populations therefore takes on heightened importance because the alternative methods (usually tagging) can be logistically and financially daunting. The details of sea-turtle life history and population structure complicate the definitions of assessment units and management units. Chapter 2 of the report reviews the current genetic methods for resolving units of assessment for sea-turtle populations (units for evaluation of status and trends) and their applications in resolving management units and strategies (units for regulation and policy that may be based on geographic location). The major challenges associated with the complex population structure of sea turtles are still being resolved so the genetic issues addressed in the report are at the forefront of conservation genetics.

A Conceptual Model of Sea-Turtle Life History

A conceptual model linking population abundance with the key demographic processes in a single coherent framework is needed because species with a long lifespan are subject to influences beyond population changes (e.g., climate, magnitude of exploitation, type of fishing effort).

The environment could change, but the population effects (in the absence of demographic information) would not be seen for a very long time if only abundance of nesting turtles were monitored. That is, the environment could become lethal to sea turtles, but the abundance data would still show no population decline so it would be difficult to interpret abundance changes and estimate population parameters accurately.

A conceptual model of loggerhead sea-turtle abundance and demography is described in Chapter 3. The model provides a simple but effective graphic device for capturing a coherent and integrated framework for the key demographic processes and anthropogenic hazards facing a sea-turtle population. The causal-loop model not only helps to identify knowledge gaps but also provides a blueprint for simulation models of sea-turtle population dynamics and for the development of population-assessment models and risk-analysis tools. It is provided as an example of what could be developed for U.S. sea-turtle populations.

Measuring Sea-Turtle Populations on Nesting Beaches and in the Water

Population sampling on nesting beaches is a valuable source of information, but authors generally do not provide detailed justifications for their data-collection techniques. Techniques for measuring abundance and other demographic parameters of sea turtles both on nesting beaches and in the water vary widely in the type of sampling, what is counted, how counts are made, and how the data are used for estimates. The techniques vary with species-specific or life-stage–specific behaviors, water depth and clarity, currents and sea state, accessibility of habitat, personnel and equipment availability, and funding. Some of the efforts use standardized methods to ensure that current datasets are compatible with older ones.

Few individual research projects are designed to collect population-wide demographic information. Most are focused on local groups of turtles and on the collection of information applied to local management issues. Other research projects collect demographic information on turtles observed or captured incidentally because of other activities, such as fisheries and power-plant operations. Thus, the location, timing, and nature of the research projects are determined by the operations that provide access to sea turtles. Variations between in-water projects notwithstanding, U.S. waters have a broadly distributed array of research that targets sea turtles. Chapter 4 reviews methods of sampling sea turtles on land and at sea and provides recommendations concerning the conditions under which they are best used and the further development of techniques.

Demographic Rates and Integrating Demographic and Abundance Estimates

Just as abundance estimates alone are not sufficient to predict or diagnose causes of population trends without estimating demographic parameters, estimates of demographic parameters without an understanding of the causes of variance and the regulating mechanisms that control them are not sufficient to understand and to mitigate adverse trends. Understanding the ecological context of demography—such key environmental mechanisms as resource availability, temperature, current systems, and oceanic productivity that influence demographic rates—is essential for an understanding of sea-turtle population status and trends. That knowledge is necessary to predict the changes in sea-turtle populations that will occur with climate change and with oceanic regime shifts that have profound effects on many critical sea-turtle habitats.

Using abundance measures for a single life-history stage can be misleading in diagnosing the status and trends of a population. Integrating abundance measures with demographic processes in a framework of modeling and data fitting provides a more robust basis for diagnosing trends, evaluating the effects of anthropogenic hazards, and defining recovery criteria. Chapters 5 and 6 review information about demography, techniques for estimating demographic parameters, some of the quantitative tools used in assessment of populations, and tools that have been applied to sea-turtle assessments; and they discuss the procedures routinely used in fishery assessments to ensure scientific rigor that could be adopted for future assessments of sea-turtle populations.

Data Management, Education, and Coordination

The fractured status and lack of coordination of sea-turtle databases are major impediments to the management and conservation of sea turtles. Throughout the United States, hundreds of projects (of varied duration) have been established to monitor sea-turtle populations or conduct research on sea-turtle biology. Projects have been conducted by federal and state agencies, universities, nongovernmental organizations, and private individuals. However, many of the data from the projects are either inaccessible or accessible only in summary formats. Consistent data collection would maximize the ability to combine and compare data among studies. Attempts have been made to standardize data-collection protocols for sea turtles, but the use of standardized protocols (e.g., description of fishing gear and operational modes, which affects estimates of incidental captures and mortality; description of handling techniques and injuries to released individuals, which affect survival estimates) is inadequate for a number of reasons. In addition, better data archives, including the storage of tissue samples, are needed.

The committee has found broad consensus among researchers studying sea turtles that the permitting process is a greater obstacle to research than is necessary for the protection of sea turtles or for meeting the requirements of the Endangered Species Act. New research projects with innovative techniques will need to be initiated to meet the data needs outlined in this report, but in numerous examples presented during committee meetings, the U.S. permitting process delayed or denied research projects, particularly when innovative techniques were involved.

THE COMMITTEE'S PRINCIPAL CONCLUSIONS AND RECOMMENDATIONS

Overarching Conclusion: Although abundance estimates are critical for assessing sea-turtle populations, demographic or vital-rate parameters are critical for understanding and predicting trends in sea-turtle populations. The committee concludes that (1) in the United States, critical vital rates have not been adequately determined; (2) the most important procedural enhancements would be improved coordination in data collection and availability, a more efficient and transparent permitting process, and increased archiving of tissue samples; and (3) sea-turtle assessments have not been isolated from broader evaluations of status and threats and have rarely included scientists from other quantitative-modeling fields.

Overarching Recommendation: The National Marine Fisheries Service (NMFS) and the U.S. Fish and Wildlife Service (USFWS) should develop a coherent national strategy for sea-turtle assessments to improve the data-collection methods, data quality, and data availability and to develop a rigorous plan for external review of data and models used to assess population status and trends. The strategy would benefit from the focused attention of expert groups that include government officials, academics, and nongovernmental organization personnel. As recommended in all expert working group documents (see Table 1.2), research should emphasize vital-rate estimation (averages, annual variance, and ecological or environmental mechanisms that drive vital rates) and improvement in abundance estimates. The most serious demographic data gaps to be addressed include in-water abundance, hatchling-cohort production, survival of immature turtles and nesting females, age at sexual maturity, breeding rates, and clutch frequency.[3] More precise estimates of anthropo-

[3] Clutch frequency refers to the number of clutches deposited by an individual turtle in a nesting season.

genic mortality are needed to evaluate impacts. All sources of data should be evaluated for quality, consistency, spatial and temporal heterogeneity and trends, and data gaps.

Detailed suggestions for improving the collection, analysis, and synthesis of data are provided at the end of each chapter of this report , and Chapter 6 describes appropriate models and procedures for assessments. Because assessments will involve different circumstances and management needs, the committee cannot recommend one standardized set of priorities for all assessments beyond its strong recommendation for a greater focus on demographic parameters. Some specific conclusions and recommendations that elaborate on the overarching conclusion and recommendation and represent the highest-priority needs are presented below.

Conclusion: Sea-turtle population assessments in the United States are based too heavily on estimates of abundance of adult females on nesting beaches. Although estimates of abundance of adult females are critical, without knowledge of accompanying changes in demographic rates at all life stages, the proximate and ultimate causes of population trends cannot be determined. Selection and evaluation of the best management options depend on an understanding of the basis of changes in population abundance.

Recommendation: NMFS and USFWS should ensure that estimates of abundance at life stages in addition to adult females are generated and that demographic rates are integrated with estimates of abundance in population assessments.

Conclusion: Inadequate information is available for population assessments because the data have not been collected or, if they have been collected, have not been analyzed or made accessible in a manner that allows them to be useful.

Recommendations:

- NMFS and USFWS should develop plans for the collection and analysis of data to address data gaps. The development should include outside experts who collect, analyze, and use the data.
- NMFS and USFWS should present a comprehensive assessment plan and a data plan to sea-turtle biologists to facilitate effective data collection for the integrated approach and to obtain input from them on improvement of the plans.
- NMFS and USFWS, with other government agencies and funding sources, should support the collection and analysis of those data.

• To avoid the overlooking of data sources, NMFS should create an on-line metadatabase[4] that identifies as many of the sea-turtle datasets in the United States and its territories as possible and that is similar to the document created for in-water projects in Florida (see Chapter 7). The database would provide information on available data, status of each dataset (e.g., computerized, hard-copy only, lost), and contact information but would not include the data. The database should be updated regularly.

• NMFS and USFWS should support a program to safeguard and make accessible as many sea-turtle databases as possible, past and present. There is some urgency to this task while data collectors are still available to provide essential metadata.

• NMFS and USFWS should partner with other government agencies, universities, and nongovernmental organizations to improve coordination among data holders. Incentives should be developed to encourage data sharing; these may include providing participating researchers with data-analysis services and data products, regional data summaries, data backup assurance, assistance with publication of results, and facilitation of collaborative relationships.

• NMFS and USFWS should arrange for a review of data now being collected under their auspices or with their support and evaluate the costs and benefits. For example, the sea-turtle stranding and salvage networks should be evaluated, perhaps with the assistance of the U.S. Geological Survey's National Wildlife Health Center.

Conclusion: Reviews of federal population assessments and research plans are not sufficiently rigorous and transparent.

Recommendations:

• NMFS and USFWS should develop a general framework for sea-turtle assessment procedures, including data evaluation, model review, and management-strategy evaluation.

• NMFS and USFWS should ensure that research plans generated in federal agencies are reviewed by panels that include federal and non-federal scientists. Using reviewers with quantitative skills, such as skills in population assessment and statistical analysis, is particularly important.

Conclusion: There are unnecessary obstacles to collection and analysis of critical data, including inadequate quantitative training of scientists and an inadequate process for issuing research permits.

---

[4] A metadatabase manages data that provide information about other data or are derived from other data.

Recommendations:

- NMFS and USFWS should partner with other government agencies and universities to improve the quantitative skills of people who are involved in designing, reviewing, and implementing the projects and assessments that are generated under a comprehensive assessment plan. These efforts will be short term (e.g., recruiting quantitatively skilled experts, improving the quantitative skills of current personnel) and long term (e.g., improving quantitative training of students).
- NMFS and USFWS should convene a working group to evaluate the permitting process for research projects and develop methods to expedite the process while meeting legislative requirements and intent. Participants should include representatives of the permitting agencies and research scientists. The review should weigh unintended consequences of permitting delays and lost research opportunities, should review the potential risks and benefits to the listed species of changing permitting requirements and procedures, and should assess the extent to which scrutiny of research permits has resulted in substantial take reductions.

1

Introduction

Long lifespans and wide-ranging migrations make sea turtles difficult to monitor and susceptible to many sources of mortality, including direct and incidental "takes" (basically any potential impact on a turtle or its behavior[1]) that occur in human activities. All six of the species that occur in U.S. waters[2]—loggerhead (*Caretta caretta*), green (*Chelonia mydas*), hawksbill (*Eretmochelys imbricate*), Kemp's ridley (*Lepidochelys kempii*), olive ridley (*Lepidochelys olivacea*), and leatherback (*Dermochelys coriacea*)—are listed as endangered or threatened under the Endangered Species Act, thereby prohibiting their direct harvest. (The seventh sea-turtle species is the flatback [*Natator depressus*], which is only found in the waters around Australia, Papua New Guinea, and Indonesia.) However, permits are available for some activities, such as shrimp fishing, dredging, and sand replenishment, that allow a specified number of incidental takes (i.e., a number of individuals that may be accidentally killed before the activity must stop). Therefore, accurate assessments are necessary to evaluate the status and trends of populations.

Regulatory decisions, such as allowing incidental takes, are best implemented with estimates of absolute population numbers, but these are unavailable because of the broad oceanic distribution of sea turtles and the very small proportion of each population that comes to land (nest-

[1] 50 CFR 17.3.

[2] U.S. waters not only refers to waters around U.S. states but also waters around U.S. territories, such as American Samoa, Puerto Rico, Northern Mariana Island, Guam, the U.S. Virgin Islands, and Palmyra Atoll.

ing adult females) (Turtle Expert Working Group, 2000, 2009). Current assessment models in the United States are based on good census data on nests and nesting females, but they lack key demographic information for extrapolating the counts to total population size accurately (e.g., Turtle Expert Working Group, 2007). With a paucity of data and of analysis of growth rates, annual survival, and reproductive frequency, current models used by the agencies can provide only relative or probabilistic numbers and trends under often limiting assumptions. For example, population A is larger than population B, population A is likely to decrease in the future, or population A is larger than it was five years ago (Heppell et al., 2003; Conant et al., 2009). Thus, they can only demonstrate population trends for segments of the population or make general predictions about the effects of disturbances on population persistence and recovery.

Because sea turtles migrate across whole ocean basins, population assessments require an international context. Global activities, such as development on nesting beaches, killing of turtles for food, and incidental capture in commercial fisheries, can contribute to sea-turtle declines and affect populations found in U.S. waters (e.g., Conant et al., 2009).

Management efforts appear to have slowed or reversed declines in some populations, such as Kemp's ridley (Turtle Expert Working Group, 2000) and Hawaiian green turtles (National Marine Fisheries Service and U.S. Fish and Wildlife Service, 2007a), but the status of many populations is still unknown or poorly understood (Table 1.1), and none have reached their recovery goals. According to the 2007 five-year status updates for each species (National Marine Fisheries Service and U.S. Fish and Wildlife Service, 2007a, b, c, d, e, f), there are many uncertainties in population structure, in productivity trends, and in the nonbreeding population of most species. However, data needed for accurate assessments of most populations are not available, prohibiting diagnostic evaluations that can benefit management.

There have been recommendations for improved data collection and analysis for status determination and assessment modeling in nearly every report and status review document published by the two federal agencies responsible for sea-turtle management—the National Marine Fisheries Service (NMFS) and the U.S. Fish and Wildlife Service (USFWS). As stated in the green turtle review, "the paucity of information regarding these [demographic] aspects continues to inhibit effective modeling of populations and prevents a full understanding of which nesting concentrations are most at risk" (National Marine Fisheries Service and U.S. Fish and Wildlife Service, 2007a). The reports repeatedly state a need for additional information on genetic relationships among nesting populations, effects of coastal and pelagic fisheries, identification of foraging areas, and identification of threats at foraging areas as key data needs for assessment

TABLE 1.1 Current Endangered Species Act Listing Status and Trends of Sea Turtles[a]

Species	Geographic Area	Listing Status	Reported Trend of Subpopulations or Nesting Aggregations			
			↑	↓	–	?
Green turtle	Florida	Endangered	1	0	0	0
	Other western Atlantic	Threatened	3	0	2	0
	Western Pacific	Threatened	2	0	1	1
	Central Pacific (U.S.)	Threatened	1	0	0	0
	Eastern Pacific	Endangered	1	0	1	0
	Eastern Pacific	Threatened	0	0	1	1
Hawksbill turtle[b]	U.S. Virgin Islands, Puerto Rico	Endangered	3	0	0	1
	Other Caribbean	Endangered	5	9	0	12
	Central Pacific (U.S. and holdings)	Endangered	1	2	0	1
	Central Pacific (other)	Endangered	0	3	0	1
Kemp's ridley turtle	Gulf of Mexico	Endangered	1	0	0	0
Olive ridley turtle	Eastern Pacific (Mexico)	Endangered	5	0	4	0
	Eastern Pacific	Threatened	1	2	1	8
Leatherback turtle	Florida, U.S. Virgin Islands, Puerto Rico	Endangered	3	0	0	0
	Other Caribbean		5	1	4	9
	Western Atlantic					
	Eastern Pacific		0	5	0	0
	Western Pacific		0	2	0	4
Loggerhead turtle	U.S. western Atlantic		1	3	1	0
	North Pacific (Japan)		3	12	0	0

[a] Listed here are the reported number of increasing (↑), decreasing (↓), stable (–), or unknown (?) subpopulations or nesting aggregations that nest in the United States or U.S. territories or that commonly occur in U.S. waters. Trends based on numbers of nests or nesting females. Data from National Marine Fisheries Service and U.S. Fish and Wildlife Service (2007a, b, c, d, e, f).

[b] Based on "recent trend" (in last 20 years).

and management. Likewise, the Turtle Expert Working Group has regularly highlighted inadequacies of its assessments for determining population size, trends (except for nesting females), maximum take levels, and evaluation of the success of various management strategies (Turtle Expert Working Group, 1998, 2000, 2007, 2009; Table 1.2). Some recent incidental take statements—required as part of an incidental take permit—have made clear how important it would be to have that information (National Marine Fisheries Service, 2005; Merrick and Haas, 2008).

TABLE 1.2 Summary of Reports by the Turtle Expert Working Group and Other Loggerhead Assessments[a]

Species	Geographic Area	Document Reference	Year	Methods
Loggerhead	North Pacific	Bolten et al.	1996	LDM[b], PBR[c], VORTEX PVA[d], RAMAS Stage PVA[e], TURTSIM[f]
Kemp's ridley	Western North Atlantic	Turtle Expert Working Group	2000	Trend analysis; LDM and LSM[g] fit to nest number
Loggerhead	Western North Atlantic	Turtle Expert Working Group	2000	Trend analysis
Loggerhead	Western North Atlantic	National Marine Fisheries Service Southeast Fisheries Science Center	2001	Trend analysis (nests); LDM

Status Conclusion	Conclusion Quotes	Recommendations
Existing incidental mortality would not have a significant effect, assuming a stable population; maximum allowable removal, 28–800, depending on life stage affected	"Although the workshop was an excellent exercise in population model integration, more research is required to further develop a suite of analytical tools robust to shortcomings in biological knowledge and data on human-caused mortality."	"Develop and implement a comprehensive quantitative framework for marine turtle recovery management including...robust procedure for monitoring turtle populations and measuring progress toward recovery goals."
Population increasing; recovery goal achievable by 2020; cannot estimate acceptable removal rates	"It is clear to the TEWG that continued work towards developing estimates of take which do not negatively impact recovery is limited in meaning without a clear understanding of the status and condition of these stocks. We are confident that future assessment teams can make progress as more data become available."	1. Obtain key vital rates, especially survival and life-stage duration. 2. Provide adequate observer coverage to statistically evaluate take throughout the species range.
South Florida stable or increasing; northern subpopulation recovery goals unlikely to be met; cannot estimate acceptable removal rates	"No method for setting strandings limits was completely satisfactory to all Group members.[h] Significant data gaps exist which limit the pursuit of complete age-specific assessments."	1. Obtain key vital rates, especially survival and life stage duration. 2. Provide adequate observer coverage to evaluate take statistically throughout the species range. 3. Define subpopulations and rates of mixing in foraging areas.
Northern subpopulation stable; Florida subpopulation increasing; 150–1,200 turtles killed in longlines each year	"It is unlikely that any loggerhead nesting subpopulation under the status quo will be extirpated over the next few years."	"It is recommended that actions to reduce juvenile mortality be identified through research and implemented as soon as feasible."

continued

TABLE 1.2 Continued

Species	Geographic Area	Document Reference	Year	Methods
Leatherback	Western North Atlantic	National Marine Fisheries Service Southeast Fisheries Science Center	2001	Trend analysis (nests)
Leatherback	Atlantic	Turtle Expert Working Group	2007	Trend analysis; Bayesian state space analysis of trends
Loggerhead	Western North Atlantic	National Marine Fisheries Service Southeast Fisheries Science Center	2009	LDM
Loggerhead	Western North Atlantic	Conant et al.	2009	DA with SQE,[i] LDM (probabilistic growth rates)[j]

Status Conclusion	Conclusion Quotes	Recommendations
Population increasing in Florida and northern Caribbean; decreasing in French Guiana; 150–530 kills in longlines annually	"While the longline fishery and the U.S. trawl fishery may not be the immediate cause in declines in nesting in French Guiana, they could be contributing to these declines."	"It is recommended that research begin immediately to identify and quantify the rate of mortality from the longline fishery, as well as mortality rates from other fisheries."
Adult population stable, increasing in some areas; 10,000–31,000 adult females, excluding unknown nesting in Africa	"Nest numbers could fluctuate considerably due to individual variance in remigration intervals, clutch number, and the reduced site fidelity in leatherbacks."	"Analyses should be interpreted with caution due to high parameter and data uncertainty; efforts should be made to develop a collaborative international research plan on population dynamics and stock structure; need to estimate demographic parameters."
1. Adult female population, 20,000–40,000+; total population highly uncertain. 2. Any reductions in mortality will improve recovery potential, but even elimination of some anthropogenic mortality sources may not be sufficient to prevent extinction.	1. "This model cannot effectively address any specific question of what the effect of mortality in a given fishery might be without making very large assumptions that are difficult to justify." 2. "Predicting future populations of loggerhead sea turtles is very uncertain due in part to large uncertainty in our knowledge of loggerhead life history."	1. Devote more time and resources to the development of improved stock assessment models of sea turtles. 2. More in-water capture–recapture and telemetry studies...to improve estimates of survival and growth.
Nine distinct population segments (DPSs) identified globally; three-fifths of DPSs with good time series show high risk of extinction; some DPSs show increasing trends, but all have possibly unsustainable anthropogenic mortality and extinction risk.	"This approach (LDM)... produced a wide range of results."	N/A

continued

TABLE 1.2 Continued

Species	Geographic Area	Document Reference	Year	Methods
Loggerhead	Western North Atlantic	Turtle Expert Working Group	2009	Nesting trend analysis; juvenile size and abundance trends

[a]This is not an exhaustive list but presents examples of the methods, conclusions, and recommendations in assessment reports.

[b] LDM = linear deterministic matrix.

[c] PBR = potential biological removal.

[d] VORTEX = individual-based stochastic simulation program for population viability analysis (PVA).

[e] RAMAS Stage = stochastic matrix PVA.

[f] TURTSIM = length-based model developed by Weatherall at the Pacific Islands Fisheries Science Center.

The fundamental theme underlying this report is that abundance assessment is essential, but abundance information alone is insufficient to diagnose the causes of trends in sea-turtle populations or to predict them. That is particularly true because abundance estimates in the United States are generally restricted to nesting females, which probably make up less than 1% of total population size (Crowder et al., 1994; Turtle Expert Working Group, 2000; Chaloupka, 2002a; Heppell et al., 2003). In addition to reliable abundance estimates of multiple segments of each population, understanding key demographic processes, such as annual survival and breeding probabilities, is essential.

WHAT IS AN ASSESSMENT?

Population assessments seek to measure the current status, evaluate trends over previous years, and predict the status of populations under various management scenarios by quantitatively evaluating population abundance and assessing such demographic parameters as productivity and survivorship (called "vital rates" that indicate the potential for change in a population). Population assessments are required when not all members of the population can be counted accurately—the case with almost

Status Conclusion	Conclusion Quotes	Recommendations
All nesting subpopulations in decline; increase in large neritic juveniles; low juvenile recruitment	"We have no time series of any demographic parameters that are appropriate to examine these hypotheses (for decline) quantitatively. We have bits and pieces of information, but lack the specific census and mortality data necessary to characterize and monitor trends."	Fundamental life-history and census data should be collected and evaluated.

[g] LSM = linear stochastic matrix.

[h] This was also the case for the 1998 Turtle Expert Working Group assessment for these species, where potential biological removal (PBR) and strandings trend analysis were suggested as methods for setting limits on strandings to trigger management action.

[i] DA with SQE = diffusion approximation with susceptibility to quasi-extinction (Snover and Heppell, 2009).

[j] Results presented as probability of population decline given current estimates of anthropogenic mortality.

all animal populations except small populations of visible, individually identifiable animals, such as California condors (*Gymnogyps californianus*), most of which are tagged. The habit of most sea-turtle species to congregate in relatively small areas ashore to deposit egg clutches (i.e., the eggs produced and laid at a single time) provides an opportunity to count animals, but these animals constitute only a small part of the total population. That feature of sea-turtle biology is shared by anadromous fish, such as salmon (family Salmonidae), which return as adults to specific spawning areas in freshwater. Like turtles, salmon except pink salmon (*Oncorhynchus gorbuscha*) have overlapping generations, but sea-turtle reproduction is more complex because their adult lifespan is long, and females do not breed every year. General life-history traits of sea turtles are provided in Box 1.1.

When more is known about a population, including age, spatial distribution, and genetics, more sophisticated models can be used for assessment wherein productivity can be evaluated for specific age groups and birth years. The value of a more sophisticated model is that, in theory, more of the uncertainty in life-history processes and vital rates can be evaluated explicitly. Ideally, a population assessment will reflect current population status and productivity accurately and can be used to

Box 1.1
Some Distinctive Features of Sea-Turtle Life Histories[a]

- Long-lived with delayed maturity (at least 10 years, maximum of 30 years or more)
- Iteroparous (nest more than once but not every year)
- Life history in marine (foraging and mating) and terrestrial (nesting) habitats
- Overlapping generations
- Undertake long migrations and disperse widely
- Nesting populations on beaches, consisting only of adult females and their eggs
- Usually deposit several egg clutches in a breeding season (the number of clutches produced by a female in a season is termed clutch frequency)

[a]Not all species have the following traits to an equal degree.

predict the effect of future management practices on the population. As in almost all marine species, population assessments of sea turtles are challenging because of a lack of critical data or a difficulty in accessing data. Box 1.2 discusses how sea-turtle assessments compare with fisheries assessments.

The term *assessment* is used somewhat generically to describe an evaluation of data to determine the status and trends of a population relative to its condition in the past or its potential condition. The results of assessments are used to address management questions, such as the maximum human-induced mortality that a population can absorb without declining substantially. Key components of the assessment procedure include independent evaluation of data quality, model suitability and robustness, and development of biologically reasonable reference points for status evaluation and management (National Research Council, 1998). A thorough population assessment needs to include a description and evaluation of change over time and space in the following areas:

- population structure (e.g., species, subspecies, distinct population segments; see Chapter 2)
- population lifecycle and demography (e.g., life stages, rates of survival, reproduction; see Chapters 3 and 5)
- population abundance and trends (e.g., evaluation and extrapolation of population indexes; see Chapter 4)
- population ecology and behavior (e.g., habitat, distribution and movements, predators and prey, disease, parasites, contaminants)

Box 1.2
Sea-Turtle Assessments and Fisheries Assessments

A population (stock)[a] assessment is an evaluation of the status and trends of a population of organisms. It is usually motivated by a concern for the effects of human activities on those organisms. NMFS has a large repository of assessment tools that have been rigorously evaluated and applied to fisheries management, but it does not have a standardized framework for data evaluation and modeling of sea-turtle populations. Similarities of some characteristics of biology, data, and management needs in sea turtles and marine fish make the application of some fisheries assessment methods to sea turtles possible, but the two kinds of populations also differ in some key respects.

With respect to biology, sea turtles are similar to long-lived, slow-growing fish, but time lags from birth to reproductive maturity (decades) are much longer than most fish. Fish with similarly long lives include some Pacific rockfish (*Sebastes* spp.), dogfish (order Squaliformes), and sturgeon (family Acipenseridae). Sea turtles are highly migratory and occur in different habitats over their lifetimes. The population structure of sea turtles is highly complex; natal homing (the process by which animals return to their birthplace to reproduce) by females creates genetically distinct nesting units, as with salmon. However, although some males exhibit natal homing, there is genetic mixing through males that have an opportunity to mate with females from different units, a pattern that is less common among fish species.

Fishery-independent data, an important part of fish-stock assessments, on turtles exist in many forms (nesting beach surveys and in-water surveys) but are not always collected with comprehensive or standardized methods that allow their incorporation into population assessments; data from many excellent sources are proprietary and unavailable for evaluation. Because available fishery data on catches of turtles are based on bycatch from more than one kind of fishery and observer coverage of many U.S. fisheries is low and of many international fisheries is absent, fishery-dependent data for estimating stock abundance, which can be important for commercial species, are not as effective for estimating turtle abundance. Finally, length distributions are available from some bycaught animals, but age distributions are not—in contrast, the ages of most fish can be determined reliably.

In fisheries management, assessment models are used to predict rates of change in biomass and productivity of a population to set harvest limits. In sea-turtle management, assessment models are used to evaluate the status of the population relative to recovery goals, to compare relative efftects of different human activities and natural stressors on populations, and to determine whether human activities that result in turtle mortality will impede recovery or increase extinction risk. Assessment of sea-turtle population status and trends is conducted according to the requirements of the Endangered Species Act and through expert working groups, recovery plan teams, and biological review teams convened by NMFS Fisheries Science Centers. Worldwide, fish-stock assessments usually are prepared by fishery agencies and—in the United States—stock-assessment teams associated with NMFS and regional management councils. Assessments

continued

Box 1.2 Continued

of fish stocks undergo rigorous review, and reports on turtle populations produced by the above-mentioned groups and teams have recently also undergone external scientific review.

---

[a] A "population" is usually defined as a group of organisms whose members interbreed and are subjected to processes that result in a common birth, mortality, and growth rate. All members of a species can potentially interbreed, and some migration occurs among populations. An example of a population of sea turtles might be all the turtles that breed on a particular beach. "Stock" (synonymous with population in this case) refers to a group with common vital rates and is often used by fisheries scientists to identify a population that they seek to manage. For a detailed discussion, see Chapter 2.

- population size (e.g., numbers of individuals, age structure, sex ratio)
- current and projected threats (e.g., human-caused injury or mortality, habitat destruction, climate change)
- sources of variability (e.g., genetic, demographic, environmental, catastrophic)

Assessments of sea-turtle populations conducted by NMFS have included all those elements but to varied degrees of detail and quantitative evaluation (Table 1.2). To be useful in decision-making, assessment requires more than simple description of trends; the large and diffuse nature of sea-turtle populations makes extrapolation of trends over time, space, and generations difficult at best and potentially misleading. Observed and potential changes in sea-turtle populations through time need to be assessed with age-structured models to determine population-wide status accurately and to diagnose causes of population change. Likewise, heuristic evaluation of possible futures under data-poor conditions has limited utility because management often requires "high-resolution" results—accurate and precise predictions of effect so that it can set take regulations and evaluate the outcomes of targeted management actions.

ASSESSMENT CASE STUDIES

To illustrate the importance of having demographic information, as well as abundance estimates in assessing sea-turtle populations, the committee briefly outlines here a comparative case study of two of the most important loggerhead sea-turtle populations in the world—the genetic stock that nests along the Atlantic coast of Florida (Ehrhart et al., 2003) and the genetic stock that nests along the Pacific coast of Australia (Limpus and Limpus, 2003a). The assessment of the Florida turtles was severely hampered by the lack of demographic information, but demographic information available on the Australian population allowed a much more thorough evaluation of hypotheses.

The loggerhead sea turtle is considered to be a globally endangered species (International Union for Conservation of Nature, 2010). It has some major nesting populations that are in decline, such as in the northwestern Atlantic (Witherington et al., 2009), and other major nesting populations that are increasing, such as in the Pacific (Chaloupka et al., 2008a) and southwestern Atlantic (Marcovaldi and Chaloupka, 2007). Increases in loggerhead nesting populations are usually attributed to conservation measures (Marcovaldi and Chaloupka, 2007), and declines are usually attributed to climate change (Chaloupka et al., 2008a) or exposure to anthropogenic hazards, such as pelagic (open ocean) fisheries (Lewison et al., 2004) or coastal fisheries (Peckham et al., 2007). But often the data that would support confidence in those attributions are lacking.

Most assessments of loggerhead sea-turtle population trends have been based on long-term monitoring of the seasonal beach nesting activity of adult females (Marcovaldi and Chaloupka, 2007; Chaloupka et al., 2008a; Witherington et al., 2009). However, monitoring only female nesting activity provides insufficient information for population assessment because adult females usually skip one or more breeding seasons, and nest counts provide no information on demographic structure because immature, adult male, and non-breeding female components are not sampled. Robust assessment of the status and trend of a loggerhead sea-turtle population suitable for population assessment and conservation-management planning requires additional information and depends on sampling of the entire demographic structure of a population resident in the foraging grounds and on deriving a range of estimates of key demographic parameter of the population.

The spatial and temporal variation in nesting activity of the northwestern Atlantic loggerhead population that nests along the Atlantic coast of Florida has been monitored for more than 20 years. These nesting populations have declined substantially over the last 10 years (Figure 1.1), but the causes remain elusive because of a lack of demographic parameters to help to diagnose the trends (Witherington et al., 2009). As a result, management agencies have not been able to predict the effectiveness of conservation strategies. A recent Turtle Expert Working Group (2009) review of the status of the loggerhead population nesting along the U.S. Atlantic coast clearly recognized that limitation: "Our ability to assess the current

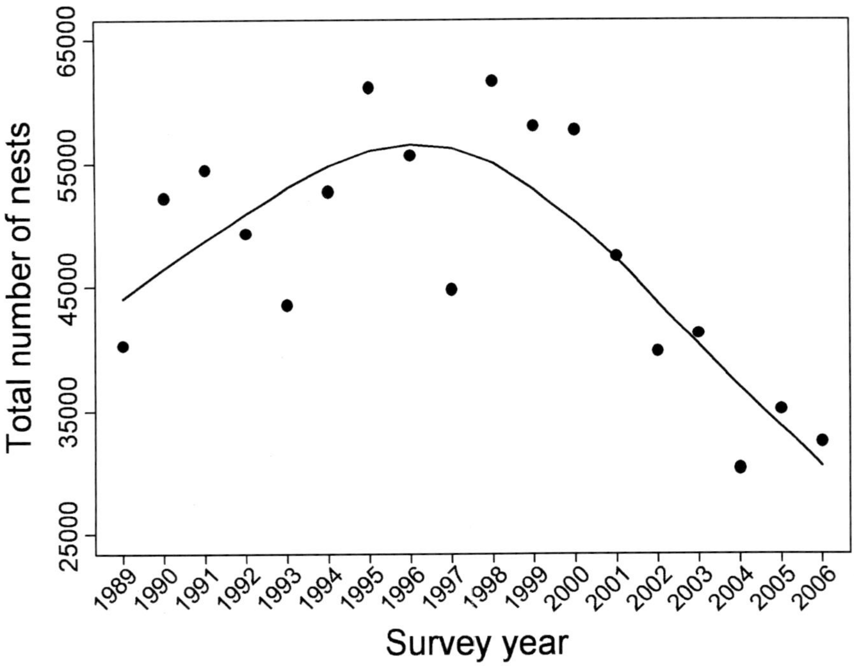

FIGURE 1.1 Annual total nest counts for loggerhead sea turtles on Florida beaches, 1989–2006. The trend line was estimated by fitting a three-knot restricted cubic spline curve to the total counts via negative binomial regression. (Reprinted from Witherington et al., 2009; with permission from Ecological Society of America.)

status of all segments of the Western North Atlantic loggerhead subpopulations is limited. We have bits and pieces of the information but lack the specific census and mortality data necessary to characterize and monitor trends for these populations." In this case, long-term abundance estimates without accompanying estimates of key demographic parameters were not sufficient to diagnose the cause(s) of the decline in nest numbers and to design suitable risk-mitigation or population-recovery strategies.

The southern Great Barrier Reef (SGBR) loggerhead is one of two Pacific populations, and much better information is available on its population trends. Loggerheads in this population nest on coral cays in the SGBR region and along the adjacent Australian mainland (Limpus and Limpus, 2003b). The SGBR loggerhead nesting population has been monitored extensively for more than 30 years (Limpus and Limpus, 2003a; Chaloupka et al., 2008a), and several foraging-habitat aggregations of the population have been extensively monitored for decades with a comprehensive capture–mark–recapture program (Chaloupka and Limpus, 2001; Limpus and Limpus, 2003a). The tagging program is coupled with laparoscopic examination of female and male loggerheads of all ageclasses residing in nearby coastal habitats. The assessments of reproductive condition support sex determination and direct estimates of breeding rates. Not only is the spatial and temporal variation in SGBR loggerhead nesting abundance well known but so are key demographic parameters, such as sex-specific and ageclass-specific survival probabilities, sex-specific breeding rates, and trends in sex-specific and ageclass-specific foraging-habitat abundance estimates (Chaloupka and Limpus, 2001, 2002; Chaloupka, 2003a).

The sex-specific and ageclass-specific abundance and demographic parameter estimates derived for the SGBR loggerhead population have provided a sound foundation for assessing the relative risks posed by exposure over the last 30 years to various anthropogenic hazards, such as coastal fisheries, pelagic fisheries, feral-animal predation of nests, coastal development effects on nesting habitat, and long-term climate change (Chaloupka, 2003a; Chaloupka et al., 2008a). For this population, it is possible to determine, for example, whether per capita fecundity (i.e., individual reproductive output) has changed, whether survival probabilities have declined, or whether the proportion of mature females has changed. It was then possible to diagnose the declining ageclass-specific abundance during the 1980s and 1990s as attributable to predation by foxes on the coastal nesting beaches and to incidental capture in coastal trawl fisheries (Chaloupka, 2003a). Both hazards have now been mitigated by federal and state government conservation agencies, and this has resulted in an apparent recovery of the stock (Chaloupka et al., 2008a). Some of the factors contributing to the ability of the SGBR loggerhead program to make those critical determinations were (1) a long-term research program maintained by a single agency with dedicated personnel, (2) a spatially extensive capture–mark–recapture program on the nesting beaches, and (3) additional capture–mark–recapture efforts in the coastal foraging habitats coupled with laparoscopy to assess both sex and breeding status (see review in Limpus and Limpus, 2003a).

The need to combine abundance trends with demographic parameters is important for all species and has been recognized for several sea-turtle species, including leatherbacks (Dutton et al., 2005), green turtles (Solow et al., 2002; Seminoff et al., 2003; Bjorndal et al., 2005), and loggerheads (Chaloupka and Limpus, 2001). These authors based their conclusions on a variety of assessments, and this committee agrees with them. For this reason, the committee has not provided a detailed review of a large number of assessments but instead has focused on methods for improving the collection of necessary data.

THE PRESENT STUDY

In light of the above concerns, NMFS requested advice from the National Research Council's Ocean Studies Board on methods for improving sea-turtle population assessments. See Box S.1 for the committee's full charge. This report is intended to help NMFS and USWFS to improve population assessments of sea turtles. NMFS is responsible for the management of sea turtles in the water, and USFWS is responsible for sea turtles on land. The shared responsibility means that cooperation between the agencies in the management of sea-turtle populations is critical. The two agencies have a history of cooperation, as in the codevelopment of recovery plans mandated by the Endangered Species Act (e.g., National Marine Fisheries Service and U.S. Fish and Wildlife Service, 2008).

The committee was asked to evaluate current and emerging population assessment techniques being applied to provide advice to managers of sea turtles in the United States. Methods for conducting population assessments vary widely from simple regression-based approaches to the use of nesting-beach trend data to more mechanistic population-dynamics models. The choice of appropriate assessment approaches depends heavily on the management question being addressed, and the ability to answer a question is often limited by the available data.

This report describes a variety of assessment types and techniques, including beach sampling, in-water surveys, genetic analyses, demographic analyses, use of bycatch (incidental take) information, and aerial surveys; reviews assessment methods; identifies information gaps; and suggests improvements for data collection. Its review of the methods used in assessments (see Table 1.2) has led the committee to conclude that most of the modeling and analysis that have been done is a valiant effort to compensate for a debilitating lack of data. Assessment methods that can be successful in fishery biology are less successful for turtles because the data generally are not as complete as they are for many commercial fish species. In addition, fishery models are focused on one main source of fish mortality—fishing—which has not been quantified for sea turtles and is only one of the anthropogenic sources of their mortality.

Filling the large gaps in the available data has far greater promise for improving sea-turtle assessments than refinement of analytical methods (Heppell et al., 2003; Turtle Expert Working Group, 2000, 2007). The committee therefore decided that its most effective approach was to focus on the data problem, and it concluded that the agencies need to do so as well. Developing a rigorous process for assessment of sea-turtle populations also has high priority. Once better data that can be evaluated in a transparent framework of scientific review are available, it will become profitable to focus more on refinement of analytical techniques.

This report does not revisit the 1990 National Research Council report *Decline of the Sea Turtles: Causes and Prevention* or any other report on the current status of sea-turtle populations or causes of sea-turtle declines. The committee felt that it was beyond its charge to discuss major stresses on sea-turtle populations, such as interactions with fisheries, and the potential effects of environmental conditions or external stresses; to detail environmental conditions or regime changes; and to assess the costs of its recommendations. Additionally, this report does not review specific assessments comprehensively, except as illustrative examples of methods and data gaps but does provide a summary of methods used. The committee was not asked to conduct its own assessments of sea-turtle populations. As a result, this report does not provide information on the status of sea-turtle populations. The committee recognizes the importance of taking an ecosystem approach to managing sea-turtle populations, but its report focuses on population assessments of a single species. Before agencies can undertake ecosystem-based approaches to assessments of sea-turtle populations, substantial information at the single-population or single-species level will be needed, as described in this report.

Because the report was prepared in response to a request from NMFS, it is directed primarily at the biologists and managers in that agency. However, the committee expects it to be useful for biologists and managers in other government agencies that have responsibilities for sea turtles and for academic and other researchers. The report also focuses on questions asked frequently of managers, on the current and emerging analyses that can be applied to address the questions, and on the sorts of data that are required to fuel these analyses.

REPORT ORGANIZATION

Chapter 2 describes the units of assessment. Typically, assessments do not cover an entire species but instead focus on populations (or stocks) or even smaller units delineated by geographic distribution or genetic information. The chapter describes the array of techniques available and in need of development for those assessments. Chapter 3 provides a conceptual model of sea-turtle life history that provides an intellectual framework for understanding survey needs and developing assessment methods. Chapter 4 focuses on methods of estimating abundance and trends in abundance and is centered on land-based and ocean-based methods. Chapter 5 discusses demographic parameters of sea turtles and what is known about them and methods and research needs. Chapter 6 discusses the importance of and methods for integrating demographic information with abundance estimates. Chapter 7 addresses a variety of issues that cut across many aspects of population assessments, including data management, education and training, the permit process, and opportunities for coordination at various levels. Chapter 8 provides the committee's major conclusions and recommendations.

2

Units of Assessment

INTRODUCTION

Understanding the units of assessment in sea-turtle populations requires clarity in the definition of nesting populations (Box 2.1). Females show affinity for specific nesting sites and thus can form subpopulations that are vulnerable to extinction. Males breed with females that can have various nesting-site affinities, thus providing male-mediated gene flow among the subpopulations. Because male gene contribution may occur on a potentially larger geographic scale, the scale defines the geographic upper limits of the nesting populations (Bowen and Karl, 2007; Lee, 2008; Wallace et al., 2009a).

The natural history of sea turtles includes several phases that are difficult to observe directly. In particular, the long generation time and oceanic habitat of juveniles are a major obstacle to studies of immature stages. The genetic identification of populations takes on heightened significance because the alternative methods (usually tagging) can be logistically and financially daunting. In this chapter, the committee reviews the current genetic methods for resolving populations as units of assessment and their applications in management units and strategies. A key theme is that population mixing in sea turtles changes with life stage, as juveniles from regional nesting populations may be well mixed, whereas breeding adults may have strong genetic divisions. Population structure also varies among genetic markers, with maternally inherited mitochondrial DNA (mtDNA) demonstrating strong population structure within breeding populations and biparentally inherited nuclear DNA (nDNA) markers

Box 2.1
Why Do Populations Matter?

Populations matter because they are the fundamental units of species management. A population is an interbreeding group that has a degree of reproductive isolation and demographic cohesiveness. Population members share key demographic features, including fecundity (i.e., individual reproductive output), sex ratio, survivorship, and recruitment. In fishery science, these populations may be called *stocks*, and in conservation, they are often termed *management units*. The terms may not be synonymous, but they all entail the key feature of reproductive and demographic independence. Isolation of populations can be spatial, temporal, or behavioral. Nesting populations may aggregate with others during periods of their life history but retain their integrity when breeding. In that case, the nesting populations form a *metapopulation*—a group of interconnected populations that have some genetic exchange (Kritzer and Sale, 2006). A metapopulation may also qualify as an evolutionarily significant unit in conservation; a distinct population segment (DPS) under the Endangered Species Act; and a regional management unit (RMU), a term developed to fit the natural history of sea turtles (Wallace et al., 2009a, b). Populations and metapopulations are also important because they are potential reservoirs of genetic diversity that retain local or regional adaptations (Jones, 2006) in that they may evolve somewhat separately, providing a source of genetic diversity that can give a species greater resilience in the face of environmental challenges.

Population genetic studies in migratory marine animals have emphasized one important lesson in the last decade: primary sampling needs to be at or as close as possible to the breeding and birthing site because these samples are not subject to dispersive life-history stages that may confound population-genetic analyses. For example, genetic surveys of bluefin tuna (*Thunnus thynnus*) across the North Atlantic yielded contradictory results but no consistent evidence of population structure. However, when young-of-year were sampled near the spawning areas, substantial evidence of population structure emerged, indicating homing by reproductive adults (Carlsson et al., 2007). Such population separation is obscured on feeding grounds by mixing of multiple populations.

sometimes showing a strong connection between breeding populations. The major challenges associated with this complex population structure are still being resolved.

Fishery scientists typically use the term *stock*, which is defined as "a discrete entity with its own origin, demographics, and fate" (Cadrin et al., 2005). It is the basic unit of management because each stock has its own unique resilience to harvest insofar as its basic vital rates (birth, death, and growth) result in a specific productivity. A stock will decline if it is subjected to mortality in excess of its ability to counteract it with new births and faster growth (density-dependent traits). Among ecologists, the term *population* is variously defined as "a group of individuals that belong to a single species and live in some defined area" (Case, 2000), "a collection of individuals that are sufficiently close geographically that they find each other and reproduce" (Akcakaya et al., 1999), and "individuals [that] form a functional unit" (Rockwood, 2006) wherein "changes are largely determined by birth and death processes" (Turchin, 2003). In essence, those definitions reflect the same concepts and are represented in the use of the term *nesting population* in the study of metapopulations. The important point is that management policies affect the timing and extent of mortality when stocks have been identified and delineated correctly, the response of the population can be estimated. In contrast, when stocks are not delineated correctly and several stocks with different vital rates are subjected mistakenly to management practices that do not account for individual population rates, the response to management is unpredictable, and smaller or less-productive stocks could become extinct inadvertently. Sea turtles may have a refuge of sorts in that they have been shown to use habitat differently depending on species, sex, and size (Hatase et al., 2002; Hawkes et al., 2006; Blumenthal et al., 2009a). However, such refuges last only until growth and reproduction induce habitat change, which may make them vulnerable.

GENETIC-ANALYSIS TECHNIQUES

The earliest analyses of genetic variation in sea turtles used protein electrophoresis (a method of analyzing the proteins present in the blood; Smith et al., 1978). That demonstrated lower genetic variability in green (*Chelonia mydas*) and loggerhead (*Caretta caretta*) turtles than in other vertebrates—a theme that would resurface in the next three decades with respect to many classes of genetic markers (e.g., Avise et al., 1992). Since the advent of direct DNA examinations in wildlife management, a number of techniques have been developed that are no longer widely used or have yet to be widely applied, such as a single-nucleotide polymorphism (a DNA-sequence variation that can occur among members of the same species). Appendix A provides a brief description and history of genetic markers and references to their use in sea-turtle studies. In the current era of genomics, the available classes of genetic markers are now known and largely well characterized. Hence, scientists expect that for at least the coming decade, the workhorse technologies for defining populations will be stable: mtDNA and microsatellites (loci at which short sequences of DNA are repeated in tandem arrays).

Mitochondrial DNA

Structure and Mode of Inheritance

The mitochondrial genome is a circular double-stranded ring of about 16,500 base pairs (bp) in turtles and most other vertebrates. This genome is housed in the mitochondria, the energy-producing organelles in the cytoplasm of eukaryotic cells. Mitochondria are inherited (with rare exceptions) through the egg cytoplasm, and sperm typically do not contribute mitochondria to the fertilized egg. This form of inheritance imparts the following two important consequences:

- Mitochondria (and mtDNA) are inherited only through the mother, providing a genetic marker for female lineages that is pertinent to sea-turtle population assessment.
- Only a single version of mtDNA is inherited. This haploid inheritance contrasts with the diploid inheritance of nDNA.[1]

An additional feature is that mtDNA accumulates mutations faster than most nuclear loci, and this makes mtDNA sequences a method of choice for microevolutionary studies, which look at small-scale changes in allele frequencies in a population.

mtDNA-sequence information was first used to test the age and isolation of the green-turtle nesting population on Ascension Island (Bowen et al., 1989) and provided genetic evidence in support of the hypothesis that females return to their natal regions to nest. The use of mtDNA-sequence data has since become a core technique for examining sea-turtle population structure.

Advantages

The reproducibility of mtDNA sequence data has been a boon to sea-turtle genetic surveys, and registries of known haplotypes[2] are maintained in the Archie Carr Center for Sea Turtle Research (University of Florida, 2001) and the National Marine Fisheries Service Southeast Fisheries Science Center (2008).

---

[1] Haploids have one complete set of chromosomes, whereas diploids have two complete sets.

[2] A haplotype is a combination of alleles at multiple loci that are transmitted on the same chromosome.

Current Use

The control region of the mtDNA is a noncoding origin of replication that accumulates mutations more rapidly than mtDNA protein-coding regions (where molecules are translated into proteins). Due to the overall dearth of genetic diversity in sea turtles (relative to other vertebrates), it is the mtDNA region of choice for population assessments. Methods of accessing the control region (via a polymerase chain reaction, a technique used to amplify pieces of DNA that generates millions of copies of a particular DNA sequence) were developed by Allard et al. (1994) and Norman et al. (1994). The resulting DNA products overlap almost completely and produce about a 400-bp fragment in green turtles (e.g., Encalada et al., 1996; Dethmers et al., 2006) and are widely used for population assessment. With improvements in DNA-sequencing technology, there is an initiative to expand the mtDNA-sequence coverage to include most of the mtDNA control region.

Limitations

The primary limitation of mtDNA-sequence data is the maternal inheritance, which precludes inquiries about many aspects of male dispersal and behavior (but see FitzSimmons et al., 1997a). A second limitation is that recently colonized nesting populations may be indistinguishable from the ancestral population even if reproductive isolation is maintained. Loggerhead nesting populations in the northwest Atlantic (northeast Florida, Georgia, South Carolina, and North Carolina) are spatially discrete but have the same mtDNA type at almost 100% frequency. Bowen et al. (1993) suggested that the paucity of genetic diversity is due to the bottleneck effect of colonization by a small number of migrants. Those coastlines were almost certainly too cold to support nesting during the last glacial epoch, which ended about 12,000 years ago so loggerhead nesting has spread northward to Virginia, the northernmost nesting site within the thermal regime for embryonic development. Although the nesting populations in northeast Florida, Georgia, South Carolina, and North Carolina are genetically indistinguishable with current mtDNA data, they are almost certainly isolated management units, based on the overall pattern of population genetic separations in loggerheads (Bowen et al., 1993; Bowen and Karl, 2007).

Microsatellites

Structure and Mode of Inheritance

Microsatellites (also known as simple sequence repeats or variable number of tandem repeats) are short segments of DNA (usually nDNA) with a repeated sequence that is 2–6 bp long. One of the most common repeats is CACACACACACACA…, which in this case would be abbreviated as CA^7 because the CA sequence appears seven times. Other versions (alleles) could be CA^5, CA^6, CA^8, CA^{15}, CA^{18}, and so on. As the numbering indicates, there can be many alleles at these highly variable loci. Like other nDNA markers, microsatellites usually have diploid inheritance, receiving one allele each from mother and father. These loci are typically scored by their mobility in a gel or polymer, which can detect fragments of DNA that differ in length by two, four, or more base pairs.

Advantages

This is another workhorse technology for sea-turtle population genetics. Because of the highly variable nature of these loci, they are used for establishing genetic relationships from family pedigrees to fine-scale population structure (Selkoe and Toonen, 2006). Microsatellites have been profitably used on most sea-turtle species to demonstrate multiple paternity. However, their broadest application may be in defining isolated populations with a biparentally inherited nuclear marker. Microsatellites, in conjunction with maternally inherited mtDNA, allow assessment of the male and female contributions to population structure.

Current Use

There are now microsatellite loci designed for each species of sea turtle; about 40 loci are available in loggerheads (Bowen et al., 2005; Shamblin et al., 2007, 2009; Monzón-Argüello et al., 2008), at least 17 in green turtles (FitzSimmons et al., 1995; Dutton and Frey, 2009), at least 24 in hawksbills (*Eretmochelys imbricata*; Lin et al., 2008; Miro-Herrans et al., 2008), at least 16 in olive ridleys (*Lepidochelys olivacea*; Aggarwal et al., 2004, 2008), four in Kemp's ridleys (*Lepidochelys kempii*; Kichler et al., 1999), four in flatback turtles (*Natator depressus*; Theissinger et al., 2009), and 15 in leatherbacks (*Dermochelys coriacea*; Crim et al., 2002; Rivalan et al., 2006a). However, these resources are even richer than the numbers indicate, inasmuch as many loci cross-amplify across sea turtles and other chelonians[3] (FitzSimmons

[3] Chelonia is the superorder that unites turtles, tortoises, and terrapins.

et al., 1995, 1997b; Jensen et al., 2006; Engstrom et al., 2007; Lin et al., 2008; Monzón-Argüello et al., 2008; Shamblin et al., 2009; Theissinger et al., 2009).

Limitations

Microsatellites are expensive to develop, requiring cloning and screening of the nuclear genome (Selkoe and Toonen, 2006). However, as noted above, research in the last decade has produced a rich library of microsatellite loci in sea turtles. A second limitation is that microsatellites are hard to reproduce between laboratories, as opposed to DNA sequences. Microsatellites are distinguished by their length differences, not by their DNA sequence, and differences of two or four base pairs may be hard to compare between laboratories. The standardization issue is a major limitation in assembling range-wide surveys with microsatellites.

METAPOPULATIONS AND MIXED STOCKS

A metapopulation is generally defined as a group of nesting populations that interact at some level but retain sufficient breeding isolation that local adaptations are maintained and vital rates can differ (Hartl and Clark, 2007). The classic metapopulation model maintains that nesting populations can go extinct because of random chance but are recolonized from other nesting populations (Levins, 1969; Hanski, 1999), maintaining the long-term stability of the species. That model probably does not apply widely to marine organisms (Kritzer and Sale, 2006). Hanski and Gilpin (1991) defined a metapopulation as a "set of local populations which interact via individuals moving among populations." That definition might apply to sea turtles in some regions, if *gametes* is substituted for *individuals*. As discussed below, sea-turtle populations, because of mating in shared feeding areas or migratory corridors, may have continuing gene flow without actually exchanging individuals.

Complex Population Structure: Life Stages

In most marine vertebrates, a survey of adults, preferably at breeding sites, is sufficient to sample the nesting population and thus to define management units. In sea turtles, because of their highly migratory nature and complex population structure, it is necessary to survey every life stage to determine the extent of connectivity among populations (Figure 2.1). Allison et al. (2003) provide a conceptual model of potential population structures, and the most common population models are reviewed in Appendix B.

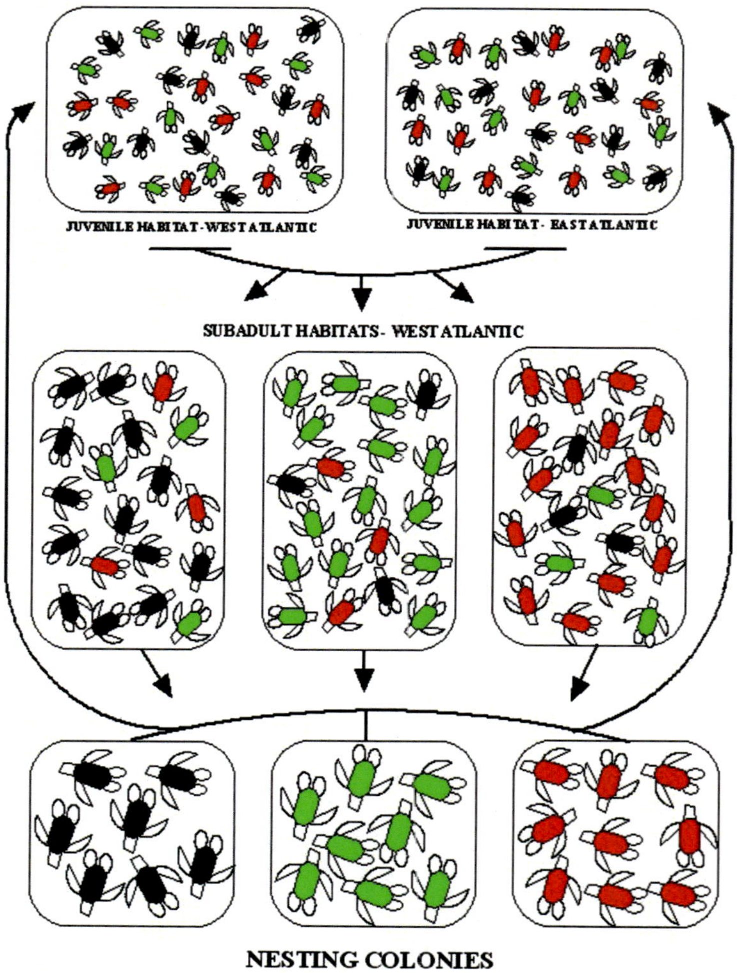

FIGURE 2.1 Population structure at three distinct life stages of the loggerhead turtle. Turtles from the three nesting populations are indicated by red, green, and black icons; the colors do not necessarily represent genetic differences. During the pelagic (open-ocean, juvenile) stage, individuals from the nesting populations intermingle in oceanic habitat, and no population structure is apparent between eastern and western Atlantic. During the subadult stage, some turtles recruit to benthic (seafloor) feeding habitat near their natal rookery (breeding population), producing low but substantial population structure (Bowen et al., 2004). During the breeding stage, females (and possibly males) show natal homing to breeding and nesting habitat, producing strong population structure (reprinted from Bowen et al., 2005; with permission of Wiley-Blackwell).

Mixed-Stock Analyses

Nesting populations separate themselves to breed, but they mingle during other life periods, such as feeding, foraging, and migration. The composition of the mixed stocks can vary (Bjorndal and Bolten, 2008). Some of them are more productive and can withstand greater mortality than others (Hilborn, 1985). Hence, when turtles mingle in shared feeding habitats, some source (nesting) populations are more vulnerable than others to common stressors. Such situations cause problems when the effects of the stressor cannot be assessed separately in each stock but only when an overall effect can be calculated.

In mixed-stock analyses, mathematical models are used to compare the genotypes (genetic profiles) of natal areas (nesting populations in the case of sea turtles) with the genotypes in feeding areas (pelagic [open ocean] or benthic [seafloor] habitats; Bolker et al., 2003). With contemporary methods, the models use maximum-likelihood or Bayesian algorithms, with the ultimate goal of estimating the contribution of each natal area to the shared feeding habitat. The methods were developed initially to estimate the contribution of salmon (riverine) breeding populations to coastal feeding populations (Grant et al., 1980). They have since been applied to sea turtles and other migratory vertebrates, and most recently methods have been developed specifically for mixed stocks of sea turtles (Bolker et al., 2003; Okuyama and Bolker, 2005).

Appropriate Applications and Current Use

One of the earliest successful studies was the assignment of pelagic juvenile loggerheads in the north Atlantic to nesting populations on the coast of North America on the basis of mtDNA-sequence comparisons (Bolten et al., 1998). The mixed-stock program SHADRACQ (Xu et al., 1994) showed that contributions from the west Atlantic nesting populations were roughly proportional to the size of the nesting populations. A similar method showed that juvenile loggerhead turtles from the west Atlantic occupy feeding habitats in the Mediterranean (Laurent et al., 1998) and that loggerhead turtles from Japan are captured in north Pacific longline fisheries (Bowen et al., 1995).

Limitations

The mixed-stock method is valuable but had low precision (wide confidence intervals) in surveys so far. A primary reason for biased (lower-bound) estimates is incomplete sampling. An unknown proportion of sea-turtle nesting is accomplished by solitary females on isolated coastlines, which are extremely difficult to sample. Nesting habitats continue to be discovered in understudied parts of the world (Yalçın-Özdilek and Sönmez, 2006; Benson et al., 2007). An additional limitation is that nesting populations are not always differentiated in haplotype frequencies. Although the precise composition of feeding populations may elude scientists in most cases, the answers provided by mixed-stock analyses are compelling when applied on an appropriate scale (Bowen and Karl, 2007) and when uncertainty in estimates is understood. Juvenile loggerheads from west Atlantic nesting populations feed in the east Atlantic (Bolten et al., 1998), and Caribbean reefs host hawksbills from multiple nesting populations (Bowen et al., 2007). On such a scale, the mixed-stock analyses can provide resolution sufficient for addressing many management concerns. However, precise estimates of the contributions of small nesting populations to feeding populations may not be possible, and this uncertainty needs to be built into predictive models used for management decisions.

DISTRIBUTION

Complex Population Structure: Female versus Male Components of Population Structure

Genetic surveys of sea turtles consistently show lower population divergence in nDNA assays than in mtDNA assays (Karl et al., 1992; FitzSimmons et al., 1997b; Roberts et al., 2004). Part of that finding can be attributed to the four-fold difference in inheritance between mtDNA and nDNA. When a zygote is formed, it has four possibilities for each nuclear locus: two from the mother and two from the father. There is only one possibility for mtDNA (from the mother). However, inheritance mechanics cannot explain the pattern completely, and part of the solution lies in differences between male and female reproductive behavior.

In the first molecular nDNA study of green turtles, Karl et al. (1992) observed lower global population structure in nDNA (Atlantic F_{ST}, 0.130; Indo-Pacific F_{ST}, 0.126) than mtDNA (Atlantic G_{ST}, 0.63; Indo-Pacific G_{ST}, 0.71; Bowen et al., 1992). (For F statistics, see the section "Analytical Techniques" below.) A reassessment with microsatellites produced the same finding (Atlantic F_{ST}, 0.038; Indo-Pacific F_{ST}, 0.024; Roberts et al., 2004). FitzSimmons et al. (1997b) reported a similar pattern for west Pacific green turtles on the basis of mtDNA and microsatellites. All three nDNA studies interpreted the pattern as evidence of substantial male-mediated gene flow between green turtle nesting populations. In other words, males apparently mate with females from more than one nesting population.

Interrookery (breeding-population) gene flow does not require departures from natal homing (the phenomenon in which animals return to their birthplace to reproduce). Overlap of feeding grounds and migratory corridors provides sufficient opportunity for mating between turtles from different nesting populations. Hence, both males and females may be homing to breeding areas near their natal beach, but gene flow can be extensive among nesting populations within an RMU (Wallace et al., 2009a). Therefore, surveys of both mtDNA and nDNA (usually microsatellites) are necessary to define populations (Bowen and Karl, 2007; Lee, 2008).

Interpretation of Genetic Data for Management

Difference between mtDNA and nDNA

The maternally inherited mtDNA data provide resolution of isolated nesting populations, but the resolution is imperfect and subject to the vagaries of each population history. For example, nesting probably expanded into higher latitudes after the last glaciation. Those new nesting populations may be isolated for hundreds or thousands of years without showing population-level differences in mtDNA sequence surveys. Therefore, it is important to focus on the overall pattern of isolation. Loggerhead turtles may show isolation on a scale of less than 100 km (Bowen et al., 2005) and green turtles on a scale of 500 km (Dethmers et al., 2006), whereas olive ridleys and leatherbacks may show high connectivity among nesting sites more than 500 km apart (Dutton et al., 1999, 2007; López-Castro and Rocha-Olivares, 2005). The overall pattern of population structure needs to be used to define management units in terms of isolated nesting populations.

The biparentally inherited nDNA (usually microsatellites) reveals the shared history of males and females and (when compared with mtDNA) shows the effect of males on gene flow and population structure. In some cases, where isolated populations do not overlap on feeding and migratory habitat, the mtDNA and nDNA data can indicate concordant population boundaries (Dutton et al., 2008). In contrast, when breeding populations overlap on feeding and migratory habitats, the nDNA can show high connectivity between local nesting populations (Bowen et al., 2005). Therefore, the nDNA should be used to define RMUs, as stated by Wallace et al. (2009a, b). RMUs may be restricted to a single isolated nesting population, as is the case for Hawaiian green turtles (Dutton et al., 2008), or may encompass several nesting populations as is the case for loggerhead turtles in the southeastern United States (Encalada et al., 1998; Bowen et al., 2005). Those RMUs are analogous to evolutionarily significant units as defined by Moritz et al. (1995) or DPSs under the Endangered Species Act (Waples, 1991, 1995).

Genetic Tags

Microsatellites can provide individual-specific genotypes (DNA fingerprints) that may serve as genetic tags to track individuals. For example, an individual genotyped on a nesting beach can be identified with high confidence from a tissue specimen taken on distant feeding habitat. When both parents are genotyped, their progeny can be assigned confidently as well; however, this application would require unrealistically high sampling of males. Genetic tags are also subject to the limitations inherent in saturation tagging (near 100% coverage of individuals), feasible for a few thousand turtles but not the tens of thousands that comprise some populations. For this reason, it may not be practical to genotype hatchlings with the expectation of matching these genotypes to turtles recaptured at later life stages. Nonetheless, genetic tags may resolve some aspects of population structure (Lee et al., 2007).

Analytical Techniques

The cornerstone of population genetic assessments has been F statistics (F_{ST}; Wright, 1943), which measure departures from random mating within and among populations on the basis of genotype frequencies. Values of F statistics generally range from zero (no population differentiation) to one (complete population differentiation). An analog that takes DNA sequence divergence into account is Φ_{ST} (Excoffier et al., 1992), usually performed in the program ARLEQUIN (Excoffier et al., 2005) or SAMOVA (Spatial Analysis of Molecular Variance; Dupanloup et al., 2002). Additional analogs are available to address the maternal inheritance of mtDNA (G_{ST}; Takahata and Palumbi, 1985); potential biases in highly polymorphic datasets (when genes exist in several allele forms), such as microsatellites (F'_{ST}; Jost, 2008); and the mutational model for microsatellites (R_{ST}; Slatkin, 1995). Many of those estimators are available from the Web service SMOGD (Software for the Measurement of Genetic Diversity; Crawford, 2009). All the genetic-distance estimators can be used to rank barriers to gene flow, as implemented in BARRIER (Manni et al., 2004).

Population genetics is a fast-moving field because of the recent development of maximum-likelihood and Bayesian approaches based on coalescence theory[4] (Kingman, 1982). Those approaches allow estimations of migration and other population parameters (Beerli and Felsenstein, 2001). Whereas F statistics provide an estimation of the number of migrants exchanged between populations, the coalescence-based approaches allow

[4] Coalescence theory uses a population sample to trace all alleles of a gene shared by all members of the population to a single ancestral copy.

some inferences about population history. It is also possible to make bidirectional estimates of gene flow with the software programs MIGRATE (Beerli and Felsenstein, 2001), IMA (Hey and Neilsen, 2007; Hey, 2010a, b), and BayesAss+ (Wilson and Rannala, 2003). That allows at least some resolution of historical sources of migrants and founders.

Historical population expansion and declines can be detected with mtDNA and microsatellites; the latter offers a more robust assessment over the timescales pertinent to population management (Beaumont, 1999). BEAST (Bayesian Evolutionary Analysis Sampling Trees; Drummond and Rambaut, 2007) and LAMARC (Likelihood Analysis with Metropolis Algorithm using Random Coalescence; Kuhner, 2009; University of Washington, 2010) are two of the most widely used programs for resolving demographic history.

Finally, assignment tests based on multilocus microsatellite genotypes may be used to assign individuals to a population of origin (Paetkau et al., 1995). That approach has the potential to detect population structure even with a high number of dispersers (5–20%; Berry et al., 2004). A related application of multilocus genotypes is to resolve population separations with patterns of genetic disequilibrium, as implemented in the program STRUCTURE (Pritchard et al., 2000; Hubisz et al., 2009). Lee et al. (2007) used assignment tests to assess population structure at the finest scale in the Ascension Island nesting population.

CONCLUSIONS AND RECOMMENDATIONS

Conclusions:

- Genetic surveys, in conjunction with tagging studies (see Chapter 4), provide the best approach for resolving the complex population structure of sea turtles.
- mtDNA surveys of nesting populations are useful for defining management units in terms of isolated reproductive populations.
- nDNA surveys are useful for resolving the male-mediated connections between nesting populations and for defining RMUs connected by nuclear gene flow. In the case of isolated regional populations, mtDNA and nDNA may indicate that management units defined with mtDNA are equivalent to RMUs defined with nDNA. RMUs may qualify as DPSs under the Endangered Species Act.
- Mixed-stock analyses can reveal the demographic links between regional nesting populations and feeding populations and can indicate which nesting populations are at risk because of habitat disturbances and fishery bycatch in feeding areas. Confidence intervals on mixed-stock estimates are usually broad, indicating problems with comprehensive sampling of turtle populations.

Recommendations:

- Researchers should examine the finest scale of female homing in each species (already underway with green, leatherback, hawksbill, ridley, and loggerhead turtles) with mtDNA surveys of nesting beaches, preferably in conjunction with tagging studies. That is necessary to resolve management units defined by female homing behavior. It requires sampling coverage of continental coastline or adjacent islands where nesting is intermittent. Adequate sample size depends on the extent of genetic diversity but may begin at about 30 per nesting population. Note that to avoid resampling the same maternal lineage specimens must come from nesting females or a single progeny per female.
- Researchers should develop a suite of at least 10–15 variable microsatellite loci for each species. That is necessary to accomplish the next three goals in population resolution and to develop individual-specific DNA fingerprints. It has been largely accomplished for sea turtles in U.S. waters with the possible exception of Kemp's ridley.
- Researchers should survey nesting populations with microsatellites to determine the extent of connectivity between local nesting populations. That is necessary to resolve the male-mediated connections between nesting populations and to resolve RMUs. Adequate sample size depends on the extent of genetic diversity (heterozygosity) but may begin at about 50–80 per location.
- Researchers should survey regional feeding populations (juveniles and adults) with mtDNA sequences to determine the source of these individuals with mixed-stock models, assignment tests, and related methods. That is necessary to determine which populations are present (and possibly at risk) in coastal and oceanic habitats. Microsatellite studies may also be useful. Priorities may be established for the most affected feeding populations.
- Researchers should survey males in breeding populations off nesting beaches with mtDNA and microsatellites to determine whether they are homing. That is necessary to resolve which populations are present (and possibly at risk) in coastal and oceanic habitats.
- Researchers should conduct a sea-turtle genome project for the explicit purpose of developing additional nuclear markers, possibly the next generation of genetic markers for sea turtles (see Appendix A). That will also provide benefits in understanding the natural history and genetic resilience of sea turtles. It may be accomplished in the context of the Genome 10K Project already under development (Genome 10K Community of Scientists, 2009).
- Researchers should develop sex-specific metapopulation models to evaluate genetic differences in dispersal. Males and females use habitat differently for feeding and reproduction, and this argues for sex-specific models for evaluating connectivity and survival. The models will increase understanding of management units and demography as outlined above.

3

Conceptual Model of Sea-Turtle Abundance and Demography

Demographic information is critical for interpreting abundance trends. Demography refers to the key vital rates or parameters, such as breeding, survival, and dispersal rates. As a concrete example, consider the common practice of assessing population status by counting nests. Setting aside sampling issues (discussed in Chapter 4), a central question in estimating the number of nests on a beach concerns the connection between variations over time in nest numbers and in population abundance. First, the number of nests on a beach in a particular year is the product of clutch frequency (the number of clutches deposited by an individual female turtle in a nesting season) and the number of females that nest on the beach in that year. To provide an index of the number of nesting females that is comparable from year to year, it is necessary either to know or have an estimate of clutch frequency or to assume that it remains constant. Otherwise, it is not possible to separate the effects on nest numbers of variations in the number of nesting females from the effects of variations in clutch frequency.

Second, the connection between the number of nesting females in a year and the number of adult females in the population is complicated by the fact that adult female sea turtles generally do not nest every year. Thus, the number of adult females in a population in a year consists of the ones that nest in that year and the ones that remain at sea. The latter number, which typically is not measured, depends on the numbers nesting in previous years, their remigration intervals (the interval between successive nesting seasons), and the survival rate of at-sea adult females. The issue is complicated by variations over time in the distribution of the remigration interval and the at-sea survival rate. Without information about remigration intervals and adult survival, it is not possible to relate the number of nesting females in a year to the total number of adult females in that year. Third, adult females make up a small part of the overall population. Their number is an index of population abundance only if their proportion in the population remains stable. Taken together, those complications in the use of nest counts as an index of population abundance underscore the importance of demographic and other information in drawing robust conclusions about a sea-turtle population from observations limited to one part of the population at one stage of the lifecycle. Hence, a conceptual model that links population abundance with the key demographic processes in a single coherent framework is needed.

CONCEPTUAL BACKGROUND

The six species of sea turtles that inhabit U.S. waters share the basic lifecycle characteristics of nesting on land with breaks of a year or more between nesting seasons and varied degrees of site fidelity (see Chapter 2), variable egg survival with an incubation period of about two months and temperature-dependent sex determination, a phase of rapid growth in the open sea, and a protracted juvenile stage of several years. The species then fall into two primary life-history groups that are based on habitat use through their lifecycle. Loggerhead (*Caretta caretta*), green (*Chelonia mydas*), hawksbill (*Eretmochelys imbricata*), and Kemp's ridley (*Lepidochelys kempii*) turtles make a developmental shift from pelagic (open ocean) to neritic (coastal, nearshore) habitat as juveniles; the discreteness of the shift may vary (McClellan et al., 2010). Leatherback (*Dermochelys coriacea*) and olive ridley (*Lepidochelys olivacea*) turtles, in contrast, remain pelagic throughout their lives. The number of years spent in preadult life stages varies among species, and lifecycle models have had some variability in the number and definition of life stages. All sea turtles undergo extensive migrations during their lives in response to changes in temperature and forage opportunities, and adult males and females migrate for mating and egg laying. With the exception of basking green turtles in Hawaii, only adult females return to land.

A simple but informative conceptual model of loggerhead abundance and demography is shown in Figure 3.1. The representation was developed for causal-loop modeling (Puccia and Levins, 1985), but it provides a generic description of sea-turtle population dynamics (Chaloupka, 2002a, 2003a, 2004) and is not tied to a particular modeling approach. This conceptual model is meant to remind the reader of the big picture and is an effective graphic device to capture in a coherent and integrated framework the key demographic processes and anthropogenic hazards facing

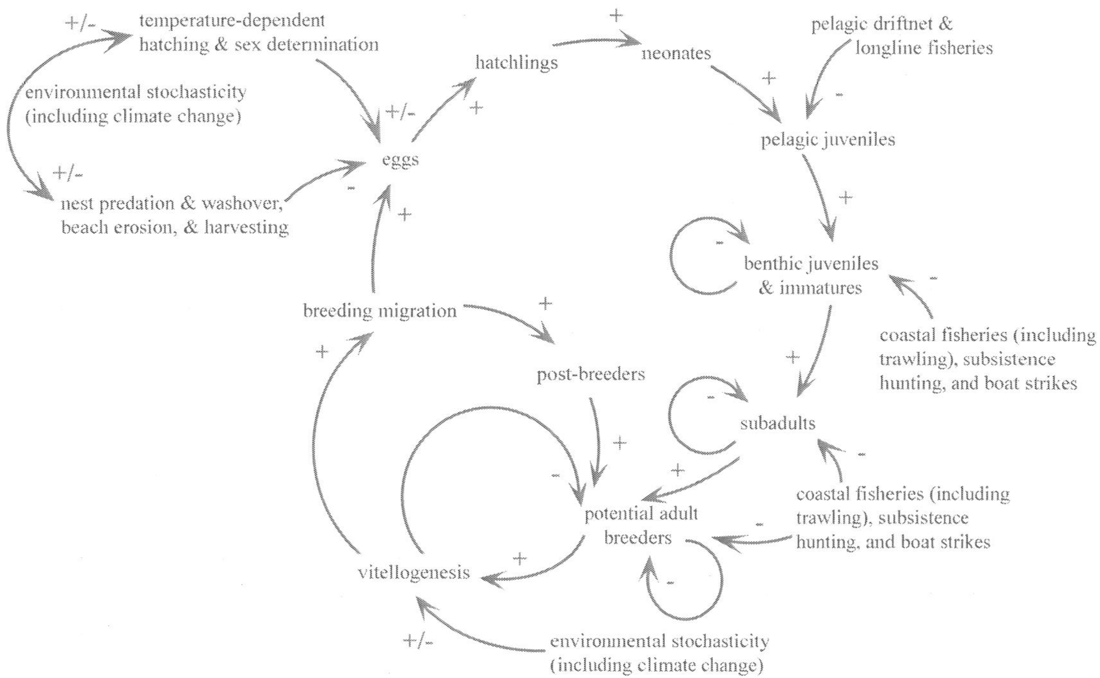

FIGURE 3.1 A conceptual or causal-loop diagram summarizing the ageclass structure and key demographic processes of the model accounting for the population dynamics of Pacific loggerhead sea-turtle populations exposed to various ageclass-specific anthropogenic hazards. +/- = causal-loop polarity: + means two components move in same direction, and - means they move in opposite directions. For instance, as more turtles breed and migrate, the number of potential breeders decreases inasmuch as females do not breed each year because of reproductive constraints. See Puccia and Levins (1985) for details on causal-loop modeling and Chaloupka (2002a, 2003a, 2004) for application to sea-turtle population modeling.

sea turtles—in this specific case, the two Pacific loggerhead populations (Bowen et al., 1994). The causal-loop model not only helps in identifying knowledge gaps but provides a blueprint for simulation models of Pacific loggerhead population dynamics and for the development of population-assessment models and risk-analysis tools. The committee presents the Pacific loggerhead model as an example of what could be developed for U.S. sea-turtle populations.

Causal-loop modeling is a special class of signed directed graph theory and is read as follows in reference to Figure 3.1. Arrowed links between variables (ageclasses and hazards) are negative if the variables change in opposite directions. For instance, as nesting-beach temperature increases above thermal maximum of embryos, egg production (the number of eggs laid in the nest) decreases because of reduced hatching rates. A positive link means that the two variables respond in the same direction. For instance, as egg production increases, the abundance of hatchlings increases. Similarly, if oceanic juvenile abundance decreases, benthic or neritic immature abundance decreases (eventually). Increasing neritic immature abundance will eventually lead to decreasing abundance as a consequence of compensatory density-dependent processes affected by per capita food supply.

Causal-loop modeling is a robust and widely used structured graphic procedure for developing conceptual models that are then used in qualitative modeling of complex biological systems (Puccia and Levins, 1985), ecosystem modeling (Loiselle et al., 2000), epidemiology (Dinno, 2007), and ecosystem-based fishery management (Dambacher et al., 2009). Causal-loop modeling also provides the basis of development of simultaneous equations or simulation modeling that is based on coupled systems of differential equations to explore ecosystem or population dynamics (Hulot et al., 2000; Chaloupka, 2003a). The qualitative conceptual models can also be embedded in probability-network models, such as Bayesian belief networks that are useful in data-poor and knowledge-vague settings (Hosak et al., 2008). A Bayesian belief network modeling approach based on the conceptual model shown in Figure 3.1 has been proposed for assessment of the relative risk posed by exposure for sea-turtle populations in Southeast Asian waters to multiple anthropogenic hazards (Chaloupka, 2007).

CONCEPTUAL MODEL FOR POPULATION ASSESSMENT

Ageclass Structure

The conceptual model of loggerhead sea-turtle population abundance and demography shown in Figure 3.1 comprises the following developmental phases or ageclasses and the abundance associated with those ageclasses (Chaloupka, 2003a):

1. Eggs are laid during the summer on sandy beaches (Kamezaki et al., 2003; Limpus and Limpus, 2003a).

2. Hatchlings emerge from the nests around two months later and escape to the sea during middle to late summer (Salmon et al., 1995).

3. Coastal hatchlings and then neonates recruit during the first year of life after escapement to the oceanic habitat (Witherington, 2002; Whelan and Wyneken, 2007).

4. Juveniles and immatures (more than one year but less than 15 years old; Chaloupka, 1998; Bjorndal et al., 2000a, 2001) inhabit productive oceanic frontal zones (Polovina et al., 2000).

5. Subadults (more than 10 years but less than 25 years old; Chaloupka and Limpus, 2001; Chaloupka, 2003a) recruit from oceanic habitat to coastal habitats and then develop into potential breeding adults.

6. Potential breeding adults (physically and physiologically mature, more than 25 years old; Chaloupka, 2003a) undergo long-distance breeding migrations (Limpus et al., 1992) to population-specific regional rookeries and culminate as courting males and nesting females.

Neritic immatures and adults are assumed to be subject to compensatory density-dependent functions; as the population density increases, the neritic component of the population is regulated by per capita food supply.

Major Demographic Processes

The major demographic processes included in the conceptual model are (1) ageclass-specific reproduction driven by environmental stochasticity,[1] (2) temperature-dependent hatching and sex determination, (3) ageclass-specific growth, and (4) ageclass-specific survival.

Reproductive Behavior

Each summer, a highly variable proportion of mature male and female Pacific loggerheads migrate from widely dispersed foraging grounds to regional rookeries in southern Japan (Kamezaki et al., 2003) or the southern Great Barrier Reef region to mate (Limpus et al., 1992). Not all females or males breed each season; a substantial fraction of the potential breeders skip one or more nesting years (Limpus et al., 1994), presumably because of variable food supply (Figure 3.2a), which can be affected by climate (Chaloupka et al., 2008a). That is one reason for the interannual fluctua-

[1] Environmental stochasticity is the variation in birth and death rates from one season to the next in response to conditions, such as weather, disease, competition, and predation.

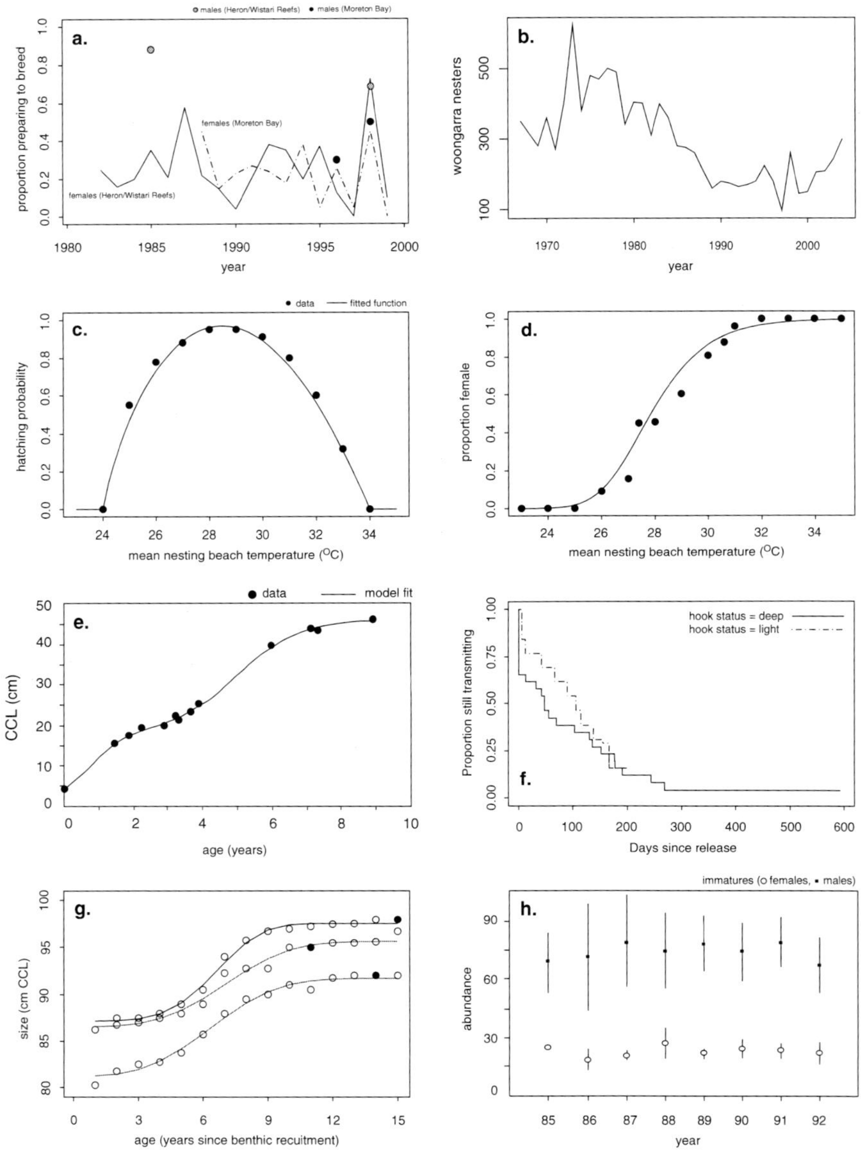

FIGURE 3.2 (a) Estimated proportion of females (curve) and males (dot) breeding each year in two loggerhead populations that make up the southwestern Pacific genetic stock (data from Limpus et al., 1994; Limpus and Limpus, 2003a; courtesy of M. Chaloupka). (b) Long-term nesting abundance (individually marked turtles) recorded at the Mon Repos rookery on the Woongarra coast, in the southern Great Barrier Reef region (from Chaloupka et al., 2008a; with permission from Elsevier).

continued

tion in the number of female loggerheads nesting each year (Figure 3.2b). It is assumed that this function is also density dependent. Assuming successful mating, the female loggerheads then lay a variable number of clutches of eggs on the sandy beaches at the rookeries over the summer nesting season.

Temperature-Dependent Hatching and Sex Determination

The probability of eggs hatching (Figure 3.2c) and the proportion of female hatchings produced are dependent on the nest temperature (Limpus et al., 1985; Matsuzawa et al., 2002). Female hatchlings are produced at higher temperatures and males predominantly at lower temperatures (Figure 3.2d), assuming that nest temperature is within the nonlethal limits for hatching (Figure 3.2c). Many southwestern Pacific loggerhead populations are female-biased because of the high summer beach temperatures at most loggerhead rookeries in the Southern Hemisphere (Limpus et al., 1994), but this is not necessarily the case with foraging-ground populations close to the regional rookery (see Figure 3.2h; Chaloupka and Limpus, 2001).

(c) Temperature-dependent loggerhead hatching probability and (d) temperature-dependent hatchling sex-determination function in Pacific loggerheads. Curve in (c) shows a Thornley type of model fit for hatching probabilities (dots); curve in (d) shows a generalized logistic-function fit for hatchling sex-determination probabilities (dots) (data for plots c and d derived from Limpus et al. [1983, 1985] for the southwestern Pacific loggerhead population; from Chaloupka, 2002b; courtesy of M. Chaloupka). (e) Pelagic loggerhead size (curved carapace length [CCL]) as a function of estimated age (from Zug et al., 1995) with the addition of estimated hatchling size. Curve shows the polyphasic logistic growth function fitted to the growth data indicated by dots (from Chaloupka, 1998; with permission from American Society of Ichthyologists and Herpetologists). (f) Expected group-specific Kaplan–Meier–Turnbull survival functions for the 40 satellite-tracked deep- and light-hooked loggerheads only without the visual clutter of the confidence intervals (from Chaloupka et al., 2004a; with permission from Inter-Research Science Center). (g) Size-at-age growth functions (open circles) for three female southwestern Pacific loggerheads recorded over a 15-year sampling period (reprinted from Chaloupka, 2003a; with permission from Smithsonian Books). Age equals years since recruitment to neritic habitat. Curve shows fitted Weibull type of growth model with AR(2) error derived in Chaloupka (2001a). Solid dots shows age at first breeding event derived from Limpus (1992, 1994). (h) Immature sex-specific Horwitz–Thompson type of abundance estimates for the loggerhead population resident on Heron reef (southern Great Barrier Reef) (from Chaloupka and Limpus, 2001; with permission from Elsevier).

Somatic Growth and Maturity

Relevant size-at-age data on Pacific loggerheads were summarized by Chaloupka (1998) for northwestern Pacific pelagic loggerheads (Figure 3.2e) that are exposed to the various hazards (Figure 3.2f). Somatic (body) growth functions have been developed by Chaloupka (2003a) for southwestern Pacific neritic female loggerheads (see Figure 3.2g). Limpus et al. (1994) and Limpus and Limpus (2003a) have shown that pelagic loggerheads recruit to a coastal or neritic habitat from the Pacific Ocean at curved carapace length (CCL) of around 70–80 cm. The pelagic phase is estimated at 10–15 years given size-at-age polyphasic growth functions (consisting of two or more phases) derived for northwestern Pacific loggerheads by Chaloupka (1998) and mark–recapture of notched southwestern Pacific loggerhead hatchlings (Limpus et al., 1994). That is a longer duration than the 6–11 years estimated for pelagic-phase duration for Atlantic loggerheads that recruit at a CCL of around 46–64 cm (Bjorndal et al., 2000a, 2003a). Once recruited to a neritic foraging ground in western Pacific coastal waters, there is apparently little evidence of either ageclass- or sex-specific dispersal behavior (Limpus and Limpus, 2001).

Size-at-age data on neritic female loggerheads were analyzed by Chaloupka (2003a) with the Weibull type of growth models (Chaloupka, 2001a) that reflect an accelerated growth phase and with longitudinal data derived for southwest Pacific loggerheads (Limpus, 1992, 1994). Some individual-based growth functions are summarized in Figure 3.2g, which indicates a neritic phase before maturity of around 10–15 years (see Chaloupka, 2003a); this is shorter than for Atlantic loggerheads that recruit at lower sizes and take longer (20 years) to mature (Bjorndal et al., 2001). The polyphasic pelagic juvenile growth function (Chaloupka, 1998) and the Weibull type of neritic-phase growth functions (Chaloupka, 2003a) suggest that Pacific female loggerheads are more than about 25–30 years old at maturity, which is consistent with estimates of the age-specific maturation period for Atlantic loggerheads (Bjorndal et al., 2000a, 2001) despite lower recruitment size for Atlantic loggerheads.

Somatic growth is negligible after the onset of maturity at a CCL of more than 90 cm (Chaloupka, 2003a). There is some evidence of sex-specific growth behavior in Pacific loggerheads (Chaloupka and Limpus, unpublished data) that is known to occur in other Pacific turtle species, such as green turtles along the Great Barrier Reef (Chaloupka et al., 2004b). Male loggerheads grow slightly faster than females at all comparable sizes in the Moreton Bay population resident in warm temperate waters. Although Pacific males might grow slightly faster than females at similar sizes, it seems that age-at-maturity is similar between the sexes inasmuch as males are also larger at maturity (Limpus and Limpus, 2003a). Somatic growth and onset of maturity may well be density dependent for loggerheads, but such an effect has been demonstrated only in a green turtle population so far (Bjorndal et al., 2000b).

Ageclass-Specific Survival

There are few reliable ageclass-specific survival probability estimates for loggerheads (see review in Chaloupka and Limpus, 2002). Loggerhead egg survival and hatching probabilities in the Pacific were based on estimates given in Limpus et al. (1985) and Matsuzawa et al. (2002). Clutch loss to tidal inundation, extreme rainfall, or beach erosion is low in the loggerhead populations in the Pacific (Limpus et al., 1985; Limpus and Limpus, 2003a). Egg predation—for example, by lizards or pigs—can be high in some southwestern Pacific loggerhead rookeries (Limpus and Limpus, 2003a) but is not a current source of egg mortality in the northwestern Pacific population. Pacific loggerhead eggs and hatchlings are also (or have historically been) exposed to numerous beach-roaming predators, such as foxes and weasels (Chaloupka, 2003a; Kamezaki et al., 2003). There are no estimates of hatchling or neonate survival after escapement to open water for the Pacific populations—some estimates of survival of about 95% during the first few hours after escapement to the open ocean have been derived for a Florida loggerhead population (Whelan and Wyneken, 2007). Bjorndal et al. (2003b) used a catch-curve approach, and Sasso and Epperly (2007) used satellite telemetry to derive estimates of oceanic juvenile annual survival probabilities of 64–81%. No such oceanic ageclass annual-survival probability estimates exist for Pacific loggerhead populations. Comprehensive estimates of ageclass-specific annual-survival probabilities for neritic immatures and adults have been derived from long-term capture–mark–recapture programs for populations from the southwestern Pacific loggerhead population, which range from around 88% to 92%, depending on whether transient behavior is accounted for (Chaloupka and Limpus, 2001, 2002). Although loggerhead annual survival probabilities are ageclass-specific, no sex-specific survival probability differences are apparent in any loggerhead population (Chaloupka and Limpus, 2002).

Anthropogenic Hazards

The conceptual model used here also provides a basis for a structured approach to risk-chain analysis, which comprises the following four major components (Merkhofer, 1987):

- Hazard (e.g., coastal trawl fisheries)
- Exposure (e.g., during the nesting season of major loggerhead rookeries along the Atlantic coast of Florida or the Carolinas)

- Effect (e.g., drowning from entanglement, failure of egg production, recruitment)
- Judgment (e.g., if exposure is extensive, loggerhead populations decline; this is considered unacceptable and warrants some mitigation strategy)

The major anthropogenic hazards to loggerhead sea turtles in general included in the conceptual demographic model are as follows (Bolten et al., 2010):

- Climate change that affects sea level and leads to beach washover and inundation of nests (Daniels et al., 2006); nesting-beach erosion (Fish et al., 2005) and temperature, which affect hatching rates (Matsuzawa et al., 2002); and hatchling sex determination (Limpus et al., 1985; Marcovaldi et al., 1997)
- Nest or emerging-hatchling predation by feral animals or natural predators attracted by human activity (Chaloupka, 2003a; Kamezaki et al., 2003; Engeman et al., 2005)
- Compaction of nesting beaches due to human activity (Kudo et al., 2003)
- Egg harvesting or poaching on mainland rookeries (Kamezaki and Matsui, 1997; Kamezaki et al., 2003)
- Nesting-female and emergent-hatchling exposure to artificial night lighting (Salmon et al., 1995)
- Hunting of nesting females or foraging-ground matures and immatures (Kamezaki and Matsui, 1997; Gardner and Nichols, 2001)
- Coastal infrastructure that affects nesting behavior and nesting-beach access (Kamezaki et al., 2003; Mazaris et al., 2009)
- Coastal development activities in foraging habitat and nesting beaches (Kamezaki et al., 2003; Limpus and Limpus, 2003b)
- Coastal fisheries (Poiner and Harris, 1996; Cheng and Chen, 1997; Julian and Beeson, 1998; Chaloupka, 2003a; Peckham et al., 2007)
- Pelagic driftnet (Wetherall et al., 1993) and longline fisheries (Polovina et al., 2000; Chaloupka et al., 2004a; Lewison et al., 2004)
- Climate change that affects food supply and hence reproductive rates (Chaloupka et al., 2008a)
- Boat strike in coastal habitats reported as a major cause of sea-turtle strandings in U.S. waters (Boulon, 2000; Chaloupka et al., 2008b)

It is assumed in the model (Figure 3.1) that neonates are not exposed to major anthropogenic hazards inasmuch as they do not appear to be caught in pelagic fisheries (Wetherall et al., 1993; Chaloupka et al., 2004a) and are not known to be caught in subsistence hunting (Gardner and

Nichols, 2001). However, ingestion of anthropogenic debris is a serious issue in this ageclass. Tar or debris was found in 20–63% and 15–17% of neonates, respectively, off the coast of Florida (Witherington, 2002). Those hazards have a direct effect on the long-term viability of a loggerhead sea-turtle population on the basis of the following key demographic metrics (Chaloupka and Limpus, 2001; Matsuzawa et al., 2002; Bjorndal et al., 2003b; Chaloupka, 2003a; Heppell et al., 2003; Limpus and Limpus, 2003a; Mazaris et al., 2005, 2006):

- Ageclass- and sex-specific foraging-ground abundance
- Nester abundance
- Ageclass- and sex-specific survival probabilities
- Ageclass- and sex-specific dispersal probabilities
- Sex-specific breeding probabilities
- Hatchling sex ratio
- Hatchling production

4

Abundance and Trends

REVIEW OF TECHNIQUES FOR MEASURING POPULATION TRENDS AT NESTING BEACHES

Techniques for collecting data on sea-turtle nesting beaches have varied in sampling approach, what is counted, and how counts are made. Authors generally do not provide detailed justifications for their data-collection techniques, but they often describe their techniques as appropriate for the existing conditions, particularly on the basis of limitations of nesting-beach access, personnel, and equipment. Historical data-collection techniques often influence current techniques. Given the variation in the range of the population covered and in whether there are data on individual turtles, it is evident that data-collection techniques are also influenced by authors' choices regarding breadth-versus-depth tradeoffs.

Types of Sampling

One-time sampling describes counts made during a short visit to a nesting rookery. Such sampling is used to determine presence and absence and approximate abundance. Data of those kinds are seldom published except when they are the only estimates early in a time series (Addison, 1997; Seminoff, 2002). One-time sampling includes serendipitous sampling based on recorded images, as in the Kemp's ridley (*Lepidochelys kempii*) nesting-female counts from the 1947 Herrera film (Carr, 1963; Hildebrand, 1963).

Reactionary sampling describes counts initiated at the onset of nesting activity. It relies on a reduced initial effort to detect when formal and more extensive efforts would result in counts being made. The most common example of reactionary sampling is counting after the recognition of an arribada (a mass nesting behavior) of ridleys (*Lepidochelys* spp.) (Valverde and Gates, 1999; Solis et al., 2008).

Systematic or periodic sampling is generally used where counts over an extensive population range or among multiple discontinuous beaches is favored over complete temporal coverage. It is used commonly for aerial nesting surveys (Hopkins-Murphy et al., 2001; Benson et al., 2007; Lauret-Stepler et al., 2007) and occasionally for ground surveys (Bjorndal et al., 1999; Sims et al., 2008). Periodicity of sampling may follow variations on a weekly schedule or may be based on tidal cycles that erase previous days' tracks (Hopkins-Murphy et al., 2001).

Sampling by index location and season allows representative locations and season dates to remain constant throughout a time series (McLachlan et al., 2006; Beggs et al., 2007; Marcovaldi and Chaloupka, 2007; Witherington et al., 2009). Although many factors contribute to the selection of index beaches and seasons, indexes are often described by authors as being representative of a population. Choices of index locations are inherently biased by logistical concerns and monitoring history. However, diversity in beach habitat (e.g., wave energy, human development), latitude, and nesting density may buffer those biases and allow representative spatial and temporal trends to be assessed (Witherington et al., 2009). Similarly, broad and consistent seasonal sampling can buffer temporal sampling biases (Witherington et al., 2009; but note the possibility of temporal shifts discussed by Weishampel et al., 2004). Sampling by index locations with variable seasons leads to uncontrolled limits on effort that affects the seasonal coverage of counts (Balazs and Chaloupka, 2004a, 2006).

A census is a count made throughout the nesting range of a population and throughout each nesting season in a time series (Witherington et al., 2009). A census also may include identification of all nesting females in a population; but in practice, researchers have accomplished censuses only for discrete island populations (Chaloupka et al., 2008c). Complete census efforts are expensive and may be unnecessary for obtaining useful measures of abundance to use in assessing trends (Jackson et al., 2008; Sims et al., 2008).

Counts

Assessment of population abundance on nesting beaches may be based on counts of eggs, tracks, nests, and nesting females. Harvested eggs have been counted as representative of reproductive effort and of nesting females on the assumption that there has been a nearly complete harvest. Such counts are seldom published except where they are the only abundance estimates early in a time series (Chan and Liew, 1996). Crawls (tracks) have been counted as representing reproductive effort and nesting females with assumptions of constant nesting success (nests and crawls) and constant clutch frequency (the number of clutches deposited by an individual turtle in a nesting season; Godley et al., 2001). Nests (clutches) have been counted as representing reproductive effort and annual nesting females with assumptions of constant clutch frequency (Beggs et al., 2007; Marcovaldi and Chaloupka, 2007; Witherington et al., 2009). Nesting females have been identified and counted as they attempted to nest (Chaloupka and Limpus, 2001; Balazs and Chaloupka, 2004a, 2006; Dutton et al., 2005; Richardson et al., 2006).

Counting Methods

Interviews to glean historical knowledge have been conducted and historical accounts reviewed to produce count data from informal assessments of nesting abundance (Marcovaldi and Marcovaldi, 1999; Meylan, 1999; Limpus et al., 2003). Those count data are seldom published except where they are the only estimates early in a time series.

Morning-after aerial surveys have recorded tracks and nests by using observers in aircraft flying on the morning after nocturnal nesting attempts (Hopkins-Murphy et al., 2001; Benson et al., 2007; Lauret-Stepler et al., 2007). That method has been used when there has been an extensive population range, discontinuous beaches, and few personnel. The aerial counts are typically calibrated to ground counts. Aerial surveys are often scheduled to correlate with tides that erase the previous day's tracks (Hopkins-Murphy et al., 2001).

Morning-after ground surveys have recorded tracks and nests by using observers on the beach on the following morning (Bjorndal et al., 1999; Marcovaldi and Chaloupka, 2007; Witherington et al., 2009). Old tracks are marked by observers on the previous day, and crawl tracks are appraised to determine species and nesting success (nests and abandoned attempts) (Schroeder and Murphy, 1999; Florida Fish and Wildlife Commission, 2007).

Counting nesting females during arribadas and other high-density nesting is one method of obtaining density estimates. In this method, sampling of turtle density on the beach is used to extrapolate the total number of nesting females (Gates et al., 1996; Valverde and Gates, 1999; Limpus et al., 2003; Solis et al., 2008). A related "stepping index" was used as a unique method for assigning turtle densities on the basis of historical accounts that described people stepping on turtles for measured distances (Limpus et al., 2003).

Tag–recapture estimates based on nesting-female encounters have been made by marking nesting females—typically with flipper tags and internal passive integrated transponder (PIT) tags—during their nesting attempts and reidentifying them as they make later nests (Chaloupka and Limpus, 2001; Balazs and Chaloupka, 2004a, 2006; Dutton et al., 2005). That method has been used to provide counts of turtles within a nesting season and to estimate total nesting females in multiple years. Temporary marks (made with paint) have been used on high-density nesting beaches where later "recapture" observations were made in waters off the nesting beach (Limpus et al., 2005). Tagging efforts on most nesting beaches involve extensive effort, typically at night; these efforts are expensive and may result in adverse effects on nesting turtles (Broderick and Godley, 1999) or other beach species, such as shorebirds (Epstein, 1999).

Modeling Counts and Abundance Estimates

Counts must be assumed to be representative if they are to apply to population abundance. Representativeness is not an issue for censuses, but most counts described as censuses take place on only a portion of a nesting population's range. Composite counts from neighboring projects based on the use of similar techniques within a population range are rare (Witherington et al., 2009), even for individual islands (Chaloupka et al., 2008c). However, reviews of nest and nesting-female counts across multiple projects have attempted to estimate population abundance on the basis of a variety of counting methods (Broderick et al., 2002).

How counts reflect abundance varies with detectability and availability of things counted and with systematic error, such as misidentification due to lost tags. At discrete sampling locations and times, estimates of nesting-female abundance are often modeled by using an observation probability function, such as a Horvitz-Thompson estimator, a general estimator for a population total, which can be used for any probability sampling plan with or without replacement (Balazs and Chaloupka, 2006), or other estimators of population totals used for varied sampling plans and encounter probabilities. The models include covariates (two or more random variables that exhibit correlated variation) that describe how available a nesting turtle is for being counted, given a specified measure of effort. In counting, effort is likely to vary within a time series because of occasional difficulties with weather, personnel, and equipment. When counts are collected as an index (standardized locations and season) and a fine spatiotemporal scale is used, missing data are filled in by using Poisson and negative binomial models (Witherington et al., 2009). Tag-loss models describe the probability of misidentifying previously counted turtles as new ones (Rivalan et al., 2005a). Although that identification error can be factored into models by using reobservation rates of nesting females, technological advances in tag persistence (e.g., PIT tags) have allowed the reduction of this error to insignificant rates.

Because counts made on nesting beaches depend on nesting activity, information on reproductive rates is required if these data are to be used for estimating the abundance of mature females. Reproductive rates often come from more completely monitored nesting beaches, but clutch frequency has recently been determined on the basis of interpretation of satellite transmitter locations (Tucker, 2010). Track counts have the greatest data requirements for estimating mature-female abundance, and counts of nesting females have the fewest data requirements. In each type of annual count, abundance estimates must account for nesting females that skip breeding seasons, which is a common trait in sea turtles. Horwitz–Thompson estimators can allow for the effect of skipped breeding on detection (Dutton et al., 2005) and have provided abundance estimates based on nesting-female counts over multiple nesting seasons. Modeling abundance on the basis of the identification of nesting females requires minimal additional data on reproductive rates because these rates can be measured as part of the method. Identification of nesting females over multiple nesting seasons can also contribute to modeling of mark–recapture rates. Open robust-design modeling using mark–recapture data has provided highly reliable nesting-female abundance estimates and detection probabilities and estimated rates of recruitment, survival, and breeding (Kendall and Bjorkland, 2001; Dutton et al., 2005; Rivalan et al., 2005b; Troëng and Chaloupka, 2007).

REVIEW OF TECHNIQUES FOR MEASURING POPULATION TRENDS IN OCEANIC AND NERITIC HABITATS

Data-collection techniques to measure abundance and other demographic characters of sea turtles in the water vary widely in many of the ways that nesting-beach techniques do. Like authors who report counts and other demographic data collected on nesting beaches, those who report similar data on sea turtles in the water seldom provide detailed justifications but often describe the techniques as appropriate for the conditions. The conditions vary with behavior that is specific to a species or life stage, water depth and clarity, currents and sea state, accessibility of habitat, availability of personnel and equipment, and funding. Some of the efforts continue with standardized methods that have been used historically to assemble comparable datasets.

Incentives to collect demographic information on sea turtles in the water influence the location, timing, and nature of data collection. Few individual research projects are designed to collect population-wide demographic information, and most are focused on local groups of turtles. Other research projects collect demographic information on turtles observed or captured incidentally because of other activities, such as fisheries and power-plant operations. Thus, the location, timing, and nature of research projects are determined by the operations that provide access to sea turtles. Personal preferences of individual researchers also have the potential to influence data-collection techniques. Their preferences may be based on opportunity, skill set, and choices regarding tradeoff between collection of fewer data on more turtles and more data on fewer turtles. In-water project variations notwithstanding, U.S. waters currently have a broadly distributed array of research targeting sea-turtle species (Eaton et al., 2008; Turtle Expert Working Group, 2009). Proceedings of a workshop on in-water sea-turtle population assessments (Bjorndal and Bolten, 2000) provide a useful introduction to application of catch per unit effort (CPUE), transect, and capture–mark–recapture (CMR) methods in these studies.

Types of Sampling

One-time sampling has been used to detect the presence and absence and to approximate the population density of sea turtles in an area, usually when there is a potential for harm from human activities, such as channel dredging or explosions (National Marine Fisheries Service, 1991; Clarke and Norman, 2005). Such counts generally apply to a time- and location-specific relative abundance or density of sea turtles although spatial or seasonal trends might be used to extrapolate results to a broader scale.

Reactionary sampling has occurred at the onset of turtle-access opportunities, such as after cold-stunning events (Witherington and Ehrhart, 1989) or other stranding episodes (Limpus and Reed, 1985a; Hart et al., 2006; Chaloupka et al., 2008b). An important characteristic of reactionary sampling is that effort is variable or not recorded regularly.

Reporting of sea-turtle observations has occurred as an element of long-term programs (as in stranding recovery) or shorter-term projects. In short-term efforts, researchers have asked boat captains, divers, or recreational fishermen to submit sea-turtle observation data (Epperly et al., 1995a; Saladin, 2007). In the social sciences, data from questionnaires and voluntary reports are subjected to extensive statistical assessments for reliability, which accompany common, but controversial, use in quantitative analyses (Manski, 1993). However, use of the reported data in sea-turtle population assessments has been largely qualitative. Reports from biologists who conduct counts of other species, typically with measured effort (James et al., 2006), could be considered as a separate category. However, counting methods and spatiotemporal distribution (over space and time) of effort are likely to be dictated by the need to detect the target species. All data that rely on reporting by second parties might be subject to underreporting (Groves et al., 1992).

Targeted opportunistic effort characterizes many sea-turtle research projects in which effort is measured and sampling locations are predetermined but sampling times are dictated by weather or other haphazard scheduling. Examples include observations or use of equipment, such as nets, that require optimal sea state or other weather conditions. Those targeted sampling efforts may occur within a framework of attempted periodic or seasonal sampling (e.g., Limpus and Reed, 1985b). They may be targeted to a broad area with haphazardly directed searches for turtles in the area. That method was chosen for small-vessel searches within an aquatic refuge with a GPS-recorded search effort (Bresette et al., 2010).

Random sampling of sea-turtle abundance is most commonly used within a stratified schedule (stratified random sampling) in which geographic groups (e.g., grid cells) are sampled independently. Stratified random sampling has been used in trawling capture of sea turtles in shelf waters (Maier et al., 2004). In those efforts, the sampling protocol of the Southeast Area Monitoring and Assessment Program (National Marine Fisheries Service Southeast Fisheries Science Center, 2001) has been used repeatedly for structuring randomized trawl samples in time and space within the southeastern United States; stations are distributed among areas where trawling is possible, and multiple species in addition to sea turtles are targeted. Fishery-observer sampling for sea-turtle bycatch has had sampling effort stratified by the timing and location of fishing effort with fishing vessels selected randomly within each stratum; the strata do not target the highest likelihood of sea-turtle bycatch (Murray, 2008, 2009) and depend on sampling locations and times chosen by vessel operators.

Many sampling efforts to count sea turtles take place at standardized index locations with periodic or haphazard scheduling. Extensive examples of those sea-turtle counting and capture efforts in the southeastern United States are discussed in Eaton et al. (2008) and Turtle Expert Working Group (2009). Authors describing sampling sites as index sites report consistently sampled representative locations chosen for high capture or observation success. Repeated sampling at the locations is often seasonal but varies between and within projects. Index locations are inherently biased by logistical concerns and monitoring history, and temporal sampling is most commonly reported to vary because of unscheduled events. In one example of continuous sampling at an index location, sea turtles are drawn into the intake water of a constantly operating power plant (Bresette et al., 1998). Although index surveys of stranded sea turtles have been proposed (Shaver and Teas, 1999), this method is not used widely for stranding counts in the United States.

Counts

Removed (killed or taken) turtles are commonly counted and are often represented in reports describing the magnitude of threats and mortality factors or in accounts of historical harvest (Witzel, 1994). Removed sea turtles include ones that are bought, sold, or transported. Parts of taken turtles are also reported, such as shell, leather, and meat.

Stranded turtles are counted as turtles that have reached land because of illness, injury, or death and that have been reported by trained observers. The U.S. Sea Turtle Stranding and Salvage Network coordinates reporting of data on those turtles in the southeastern United States and on U.S. islands in the Caribbean Sea (Southeast Fisheries Science Center, 2010). The data are used most commonly in qualitative assessments of abundance (e.g., to detect periodicity in mortality events; Crowder et al., 1995) and generally are presented as a combined function of relative abundance and relative mortality (or morbidity). The data also have been used in conjunction with counts of nesting turtles in the same region to estimate mortality (Epperly et al., 1996). In addition to superimposed effects of abundance and mortality, stranding counts are influenced by physical oceanographic factors, including winds, currents, and temperature (Epperly et al., 1996; Hart et al., 2006). Trends in stranding counts vary with observation and reporting effort (Tomás et al., 2008). Collection of data on stranded sea turtles is discussed in more detail in Chapter 5.

Captured sea turtles are counted either as turtles obtained through targeted efforts or as turtles captured incidentally. Capture of turtles allows researchers to collect data in addition to simple counts and to mark released turtles with tags that identify and track. The additional data allow counts to be divided by categories, such as size, sex, and genetic origin. Tagging, release, and recapture of identified turtles facilitate estimation of abundance and survivorship and allow studies of behavior and physiology (Bjorndal et al., 2003c, 2005).

Observed turtles have been counted from underwater (Leon and Diez, 1999) and from vessels, land, or air and include turtles recorded both at and below the water's surface with varied associated information. These observations have a higher encounter rate per unit effort but have lower information return per encounter than turtle captures. Occasionally, observation counts are made in conjunction with sea-turtle capture efforts (Bresette et al., 2010).

Counting Methods

Sea-turtle abundance has been estimated from interviews (Epperly et al., 1995b; Meylan 1999), historical accounts (Witzel, 1994; Jackson et al., 2001), and archeological data (McClenachan et al., 2006; Allen, 2007). The data are often the only representations of abundance early in a time series. Because of uncertainty in how reports and extrapolations are related to actual abundance, little analysis of these data has been conducted. Some counts are discussed in terms of orders of magnitude of abundance, and harvest data are most commonly presented without measures of associated effort.

Aerial surveys (Kenney and Shoop, in press), vessel surveys (Bresette et al., 2010), and diver surveys (Makowski et al., 2005) of sea turtles are conducted along transects and vary in two important ways: in their geographic scope and in the associated data that allow extrapolation of observations to estimations of turtle density and abundance. Aerial surveys have the largest geographic scope, but there are presumed tradeoffs in low detectability and misidentification of species, especially when flight speeds and altitudes favor marine-mammal target species rather than sea turtles (Marsh and Sinclair, 1989). Most surveys use a variation of line- or strip-transect methods to estimate relative density and abundance from observations. Some surveys are conducted in conjunction with measurements of turtles' surface time so that an availability function can be used to estimate absolute density and abundance (Mansfield, 2006).

Aerial and vessel surveys of sea turtles can vary in objectives, methods, and operating models, and their spatial extent can range from tens to thousands of square kilometers. Since the first ones with light aircraft, most large-scale surveys have applied line-transect theory used for population assessment of marine mammals (Buckland et al., 1993, 2004). The Cetacean and Turtle Assessment Program works to detect seasonal (quarterly) patterns and habitat use and covers about 280,000 km^2 of the northeastern U.S. continental shelf (Shoop and Kenney, 1992). Similarly, in the Gulf of Mexico region, a series of separate geographic blocks were surveyed to portray seasonal distribution and abundance patterns (Fritts et al., 1983). In the southeastern United States, large-scale aerial surveys were used to detect sea turtles from North Carolina to the Florida Keys (e.g., Schroeder and Thompson, 1987); others were conducted in juvenile or estuarine habitats, such as the Carolinas (Braun and Epperly, 1995) and Chesapeake Bay (Musick et al., 1994). Although sea turtles are included and counted in the long-running Southeast Right Whale Survey coordinated by several states, the National Marine Fisheries Service (NMFS), and the New England Aquarium (Slay et al., 2002), the sea-turtle sightings data were not used for assessment purposes. In a detailed review, Kenney and Shoop (in press) present aerial-survey design, sampling limitations, and objectives of specific surveys conducted in the United States and abroad.

The challenges in detecting sea turtles are similar to those in detecting small marine mammals, including glare, sea state, field of view, observer fatigue, and similarity of appearance. Species identification of sea turtles is difficult, even for well-trained, highly experienced observers (e.g., Marsh and Saalfeld, 1989; Henwood and Epperly, 1999). Turtles smaller than 60 cm in carapace length are difficult to detect from fixed-wing aircraft flying at any altitude or speed although smaller ones (25–30 cm) may be identified correctly from airships (lighter-than-air craft; Kenney and Shoop, in press). Research design for aerial assessment of sea turtles is complex, and the surveys are expensive. Estimation of density or absolute abundance presents a number of sensitivity issues (Burnham et al., 1985; Gerrodette, 2000).

When surfacing behavior must be considered, a correction factor is used for unobserved animals, but sea turtles' dive patterns vary with size, species, ambient temperature, and activity (Lutcavage and Lutz, 1997). If a number of species are present in an area, a single correction factor for submerged (undetected) turtles could be highly biased. A major challenge is the assumption that animals are randomly distributed and can be equally sampled; abundance surveys might be designed to represent expected densities in different habitats.

Novel imaging methods developed for other fields of study have the potential for use in aerial and vessel surveys of sea turtles. They would allow both an increase in the proportion of turtles available to be counted and an increase in the recording of observed turtles in a way that would reduce detection bias. For example, vessel-mounted multibeam sonar is in use and allows imaging of individual fish within schools. The signal resolution of some systems is sufficient to estimate sizes of individual animals in decimeters at a distance of 90 m from the vessel (Lutcavage et al., unpublished data). Laser-based ranging systems (using light detection and ranging [LIDAR]) and radar-based ranging systems have also been used to detect marine animals and to image fish schools and, in principle, could detect turtles within the sampled swath (Hunter and Churnside, 1995; Brill and Lutcavage, 2001) although high costs and expensive post-processing activities have limited its use. High-resolution video and still photography coupled with attitude sensors that enable spatial referencing or georeferencing (assigning geographic coordinates to an image) are new approaches that might be used in aerial surveys. These imaging techniques allow recording of observations and estimation of size of detected targets and could be combined with computer software "trained" to recognize species differences that cannot be discerned by human observers. Integration of new technology and engineering solutions might help to overcome the current limitations of aerial surveys, namely species identification, size estimation, and presence of submerged animals. Coupled with species-specific understanding of dispersal rates, vertical behavior, and environmental associations from data-logging tagging studies, direct aerial or in-water surveys may lead to better indexes or absolute estimates of regional abundance.

Sea-turtle capture methods vary fundamentally by whether they have a measurable associated effort and whether sea turtles are the targeted species. Targeted sea-turtle capture methods with effort measured by net-soak time, tow time, and net size include use of tangle nets and trawl nets (Ehrhart and Ogren, 1999). Other targeted capture methods with variable potential for measures of effort include use of hand capture (Limpus and Reed, 1985b; Bjorndal et al., 2005; Bresette et al., 2010), dip nets (Witherington, 2002), hoop nets (Beavers and Cassano, 1996, James and Mrosovsky, 2004), and strike nets (Ehrhart and Ogren, 1999). However, some researchers have used measured effort associated with initial observation of turtles that are later captured by these methods (Leon and Diez, 1999; Witherington, 2002; Bresette et al., 2010).

Incidental capture of sea turtles may have either a measured or an uncertain effort associated with turtle captures. Captures from fisheries—including use of pound nets, trawls, gill nets, seine nets, longline hooks, and rod and reel—have various levels of recorded effort that depend on cooperation and communication with fishermen. In some cases, close relationships between researchers and fishermen allow high certainty of effort measurement (Epperly et al., 2007). In the case of power-plant entrapment, effort is measurable in terms of water flow and is constant except for occasional outages (Bresette et al., 1998).

Modeling Counts and Abundance Estimates

Data representing observed turtles are applied most often to measures of relative abundance or density by using point-count methods, strip-transect methods (Marsh and Saalfeld, 1989), or, more commonly, line-transect methods (Epperly et al., 1995b; Beavers and Ramsey, 1998), each with assumptions regarding detectability and availability (Buckland et al., 1993). Point-count methods are generally thought of as methods to approximate indexes of relative abundance and are not commonly used to estimate abundance or density. Although they have an assumption of constant proportionality between observation periods (a constant probability of detection), the methods do not allow the assumption to be tested.

The best example of modeling estimates of relative abundance on the basis of transect observations is the use of distance methods (Buckland et al., 2001; Eguchi and Gerrodette, 2009), in which observers measure the distance to each observed animal. With these methods, it is possible to model detectability of subjects and their density by using observed distances and counts, and researchers model the reduction in detection probability with distance from a transect, assuming perfect detectability on the line itself, or specify an effective strip width that includes a high proportion of observed animals. In the recent decades, there has been sufficient development of line-transect and strip-transect approaches, and a substantial body of peer-reviewed and technical literature addresses theory, analytical assumptions, and practical applications of survey design for in-water studies. Assumptions of line-transect versus strip-transect theory dictate survey protocols and sampling design, and reviews have concluded that line transects are preferred because they require fewer assumptions about detectability and use all the sightings in the analysis (Burnham et al., 1985; Marsh and Sinclair, 1989; Kenney and Shoop, in press).

CPUE is a measure of relative abundance that may involve removal of turtles from the population and may be applied in a variety of ways, including intentional capture for research and bycatch from fisheries. However, CPUE does not always have a linear relationship with density (Hilborn and Walters, 1992). Fishery studies have shown that the nonlinear function is most common in circumstances in which sea turtles are typically captured by individual research projects, namely captures of clustered animals with effort concentrated in a small spatial scale where turtles are most abundant. Sample biases, inconsistent methods between projects, and low and variable capture rates can make it difficult to justify the use of CPUE as a quantitative index of abundance statistically. However, pooling of regional capture efforts may reduce the difficulty. Within a capture project, reducing sampling bias would rely on standardization of sampling season, capture gear, and other methods that affect capture efficiency. Ideally, sampling would be randomized in space and time, especially if CPUE is to apply regionally. However, nonrandom sampling, as at individual index sites, can be valuable in assessing qualitative annual trends. The problems with the reliability of CPUE to represent relative population abundance are likely to be reduced as multiple capture projects are used within a regional meta-analysis. Although, unlike regional aerial surveys, a multiproject CPUE analysis would still rely on discrete sampling points; benefits over aerial observations would include positive species identifications and separation by sex, genetic population, and size (age).

CMR estimates of abundance are possible wherever sea turtles are captured by any method, with or without measured capture effort, as long as recapture rates are high enough (Le Gall et al., 1986; Chaloupka and Limpus, 2001). CMR also includes marking (e.g., painting) and resighting of turtles, which would not involve recapture. In addition to abundance estimates, captures and CMR modeling allow assessment of information on demographic structure and survivorship rates. Pine et al. (2003) offer a review of study designs that use CMR under a variety of assumptions and information needs.

As with CPUE from individual-capture project locations, CMR can estimate regional population abundance more powerfully if it uses multiple capture sites. CMR data collection coordinated within a networked array of sites, including nesting beaches, would provide one of the most detailed and powerful datasets possible for assessments of sea-turtle abundance and for measurement of many important demographic rates (Chaloupka and Limpus, 2001; Bjorndal et al., 2005). Wider networking of capture sites allows a wider inclusion of turtles' state variables, such as sex, genetic identity, size, physiological condition, breeding status, and geographic location.

Integrative Methods

In many cases, aerial surveys are undertaken to assess a variety of air-breathing species (e.g., Marsh and Sinclair, 1989; Palka, 2000), but their distributions and dispersal patterns may not be similar. That is especially true in foraging areas because sea birds, mammals, and sea turtles target different prey and would tend to aggregate where their food is concentrated. How well in-water surveys represent true abundance is never known, but surveys that use existing knowledge of sea-turtle dispersal rates, vertical behavior, and environmental associations better are needed. Sonic tracking and satellite telemetry can be used to provide context for interpreting surface abundance patterns and linkage between study areas (Blumenthal et al., 2006). Integrative studies that use different technologies are being applied to large pelagic fish and sharks (see Nielsen et al., 2009). For example, are habitats being used primarily for feeding, refuge, transit, nesting, or mating? Novel sensors that record behavior—such as orientation magnetometers or "daily diary" tags (e.g., Wilson et al., 2008), mouth sensors (Hochscheid et al., 2005; Myers and Hays, 2006; Fossette et al., 2008), and stomach-temperature "pills" developed by Southwood and Kirby (2008)—can detect foraging events, and GPS-satellite–linked tags provide high-resolution locations where events occur. Various behaviors have been monitored with animal-borne imaging systems (e.g., critter cams; Heithaus et al., 2002; Reina et al., 2005; Seminoff et al., 2006; Arthur et al., 2007). Acoustic arrays, video monitors, and tracking networks now deployed primarily to track marine mammals or fish species may be used to monitor behavior of sea turtles in a variety of habitats or "hotspots." Broad-scale deployment of acoustic receiver systems, such as the Ocean Tracking Network (O'Dor and Stokesbury, 2009), establishes the potential to integrate information on sea-turtle movements across state and national boundaries.

Integrated spatial and temporal information on dispersal behavior is necessary to understand and inform interpretation of abundance patterns obtained with aerial or in-water methods. In addition, oceanographic, remote-sensing, and climactic information (e.g., presence or strength of El Niño, Gulf Stream eddies, tropical depressions) provide additional context for understanding abundance patterns (Saba et al., 2008; Mansfield et al., 2009a).

In ecosystem approaches to marine-resource management, there is a new emphasis on fishery-independent surveys to provide better assessment tools and understanding (Cotter et al., 2004, 2009; Jennings, 2005). Some of the approaches include the development of indicator series of survey-based models (Rice and Rochet, 2005), which may offer good applications for sea-turtle assessment, that by tradition lack CPUE-based frameworks.

CONCLUSIONS AND RECOMMENDATIONS

Measuring Population Trends on Nesting Beaches

Conclusions:

- Choice of techniques to estimate adult-female abundance on nesting beaches has been influenced by logistics, personnel availability, opportunity, existing networks, and historical data. Few studies have sought to optimize the information gathered, given resource expenditure.
- Most U.S. nesting beaches have programs in place to count nests as a measure of sea-turtle abundance. The programs have extensive geographic coverage but do not provide direct turtle counts, measure recruitment, or estimate adult-female survival and reproductive rates. Few programs measure representative egg-to-hatchling survival.
- Multiannual near-saturation tagging of nesting females on the nesting beach provides a straightforward way to count turtles, measure recruitment, and estimate survival and reproductive rates, but the required effort is extensive and would be difficult and expensive to maintain throughout a population's range and nesting season for a statistically powerful time series.
- Seasonal nest counts require less effort per spatiotemporal unit. However, these counts estimate adult females indirectly (with associated error) and do not produce other information on vital rates.
- Interpretation of tracking data to measure reproductive rates has been used as a substitute for direct identification of large numbers of nesting females through tagging studies.

Recommendations:

- NMFS and the U.S. Fish and Wildlife Service (USFWS) should work with the states, and with other countries, to coordinate existing nesting-beach data collection so that effort is balanced between geographic scope and depth of information gathered.
- Agencies should facilitate a tiered method of nesting-female abundance counts on beaches spanning a spectrum of data scope (breadth and depth proportions). An example of such a tiered method is (1) standardized population-wide track or nest counts with spatiotemporal sampling that could detect biologically significant spatial trends; (2) nest counts in representative index locations and seasons with spatiotemporal sampling over a time series long enough to detect biologically significant spatial and annual trends (e.g., a change of 1% per year); and (3) near-saturation identification tagging in representative index locations and seasons with mark-and-recapture rates of sufficient statistical power to detect biologically significant changes in annual number of nesting females, breeding rates, recruitment, and survivorship.
- The proposed methodological tiers ideally would be divided among existing research and conservation efforts and groups. For example, beach surveyor networks coordinated by government, nonprofit, and university-organized entities, are effective in maintaining broad-scale track and nest counts for long time series. Those groups may also coordinate indexed nest counts and conduct near-saturation tagging efforts. However, extensive tagging programs may be attractive to individual researchers in consulting firms and universities because of the potential that such projects have for ancillary basic and applied research.
- Because existing datasets and data-collection networks are important in planning efforts to measure nesting-female abundance on beaches, attention should be given to coordination and training that would focus existing data collection on statistically valid and powerful sampling and methods, measurement of observational error, and the recording effort.
- NMFS, USFWS, and the states should facilitate representative sampling of nesting females tracked with satellite tags, GSM telephone tags, or other technologies to describe clutch frequency and test hypotheses on nesting-site fidelity. Those methods have a lower potential to generate survival rates than extensive marking with PIT and flipper identification tags. However, those tracking methods are useful for estimating clutch frequency in populations that nest over a broad geographic range where the mark–recaptures rate per unit effort is low. Remote tracking efforts that take place in conjunction with extensive marking of nesting turtles are recommended as a powerful combination of comparative methods.

Measuring Population Trends in the Water

Conclusions:

- Given an extensive distribution of current studies of sea turtles in the water, there is the potential for an integrated network of sampling projects to assess abundance and trends on local and regional scales.
- This integrated network would comprise intensive, low-variance measures of relative or absolute abundance in multiple, turtle-dense areas (i.e., index sites) and less-intensive, broad-scale measures of relative abundance throughout the same regions. Index sites may need to be geographically broad where turtle densities are determined by transient oceanographic features.
- Establishment and coordination of an integrated network, participant training, data sharing, and effective data management will require NMFS to provide resources, such as specialized program funding, expertise, and adequate staff.
- Assessments of relative abundance are sufficient for determination of trends; however, localized measures of absolute abundance are helpful in evaluating incidental catch and mortality and other takes.
- CMR efforts in various international locations have contributed to local and regional analyses using open robust design models to estimate relative or absolute abundance.
- Less-intensive, broad-scale measures of regional relative abundance (e.g., aerial surveys) are not a substitute for abundance measures in index sites. However, broad-scale surveys can fit into an integrated network of sampling projects by calibrating counts between well-sampled index sites and poorly sampled sites, by identifying spatial overlap with fisheries and other human activities, and by providing the only possible measure of relative abundance in inaccessible areas.
- Broad-scale measures, such as aerial surveys, may not be appropriate for estimates of regional abundance because of costs associated with long-term sampling and maintenance of extended synoptic surveys. They are most useful when coupled with measures of detectability and availability that allow estimation of turtle density.
- Measures of relative abundance based on aerial surveys will become more useful when detectability is improved by application of new technologies (e.g., LIDAR, multibeam sonar) and collection of more detailed information that would allow abundance to be assigned to specific size or ageclasses of a population's conceptual model. For example, new instrumentation, such as image mosaic and rectification, will allow accurate size assessment and help to define relationships and demographic overlap of surveyed areas and index sites where turtle life stages and genetic stocks are known.

• Fishery observer data can contribute to relative-abundance estimation when effort and vulnerability to capture (or detection) is understood (how it varies with catch rate) and when information that would allow abundance to be assigned to ageclasses of a population's conceptual model is collected.

Recommendations:

• NMFS should play a leadership role in assessments of sea-turtle abundance and trends by funding and coordinating an integrated network of sampling projects.

• Index sites should have internal (within-project) consistency in methods. Methods should be standardized between sites with similar sampling conditions but need not be standardized among all index sites.

• Random or periodic sampling in index sites is recommended to reduce sampling bias; however, consistency in bias should allow determination of representative trends in relative abundance.

• Index sites should be representative of geographic areas, genetic stocks, and life stages.

• Effective coordination should include training participants in network protocols and data reporting, application of incentives, and stipulation of requirements to achieve data sharing.

• Effective data management should include open access to data, metadata, and data products and facilitation of analyses by third parties.

• To improve its program for assessing abundance and trends, NMFS should develop a networked array of sites, having long-term CMR efforts that would support local and regional analyses with open robust-design models to estimate relative or absolute abundance specific to ageclasses in the conceptual models of populations. Assigning abundance to a conceptual model implies that turtles are identified by their genetic stock and that abundance measures apply to specific life stages. Secondarily recommended for multiple index sites are measures of relative abundance with quantified effort and estimated values for detectability, having relative-abundance measures that can be assigned to specific ageclasses of a population's conceptual model. This includes most in-water capture studies with quantified effort.

5

Demographic Rates

Knowledge of demographic rates and trends are essential for accurate assessments of sea-turtle populations, as outlined in the discussion of the conceptual model in Chapter 3. The reasons for changes in sea-turtle abundance cannot be diagnosed—nor can management plans to mitigate declines in populations be developed—without demographic knowledge. This chapter introduces the various demographic parameters and methods for generating estimates of them. Chapter 6 describes applications of the different demographic parameters further.

All demographic parameters exhibit variation within and among species and populations and over space and time; some—such as clutch frequency (i.e., the number of clutches deposited by an individual turtle in a nesting season), interbreeding intervals, and somatic growth rates—vary within individuals over time. To develop an accurate assessment, such data need to be collected on all populations, on large spatial scales, and over many years. Caution is needed when extrapolating estimates between species and populations and even within populations for different years and habitats. However, estimation that accounts for variation is expensive. Methods to estimate demographic parameters at reasonable cost are needed so that they can be monitored frequently to detect changes. Moreover, estimation of variance about the mean, not just point estimates, is critical.

The ecological context of demography—that is, the key environmental mechanisms that regulate demographic rates, such as resource availability, temperature, current systems, and oceanic productivity—is necessary for understanding sea-turtle population status and trends fully. That knowledge is critical for predicting the changes in sea-turtle populations that will occur with climate change and with oceanic regime shifts that have profound effects on many important sea-turtle habitats.

Demographic parameters are not of equivalent value for diagnosing status and trends in populations. Some vital rates are influenced more than others by environmental factors—probably acting largely through nutrition. For example, nutrition affects age at sexual maturity, clutch frequency, and the number of years between breeding seasons, but it does not affect clutch size (Bjorndal, 1985). In populations with ample high-quality food, somatic growth rates, body condition, and clutch frequency will be high, and interbreeding intervals will be small. Populations that have poor food resources or that are approaching carrying capacity, at which competition for food is high, will exhibit the opposite.

BREEDING RATES AND ADULT-RECRUITMENT PROBABILITIES

In most species of sea turtles, females generally do not reproduce in consecutive years but at variable intervals of two years or more. The probability that a female will reproduce in any given year (breeding rate) is affected by nutrition (Bjorndal, 1985), environmental factors, and migration distance between foraging grounds and nesting beaches (Limpus and Nicholls, 2000; Solow et al., 2002; Troëng and Chaloupka, 2007). Knowledge of breeding rates is critical for understanding the highly variable numbers of clutches deposited in successive years on nesting beaches (Hays, 2000; Broderick et al., 2001; Solow et al., 2002) and for interpreting population trends.

Estimates of breeding rates of females have been derived from mark–recapture studies on nesting beaches using an "open robust design"—a specific mark–recapture method—with hawksbills (*Eretmochelys imbricata*; Kendall and Bjorkland, 2001) and leatherbacks (*Dermochelys coriacea*; Dutton et al., 2005). Mean remigration interval (the number of years between successive breeding seasons) has been estimated more commonly in sea-turtle studies and approximates the inverse of breeding rate. Although not as useful as breeding rate for demographic models, the remigration interval does offer important insights into the productivity of the population and population density relative to carrying capacity (Saba et al., 2007; Troëng and Chaloupka, 2007). Remigration interval is usually measured as the number of years that elapse between sightings of individual tagged females at a nesting beach. Thus, values are biased to shorter intervals because of tag loss and human-induced mortality in that the probability of both factors increases with the length of the remigration interval. Values are biased to longer intervals when incomplete sampling on the nesting beach results in females being missed in intervening breeding seasons.

Breeding rates of male sea turtles have been poorly studied, and more information is needed. Males may breed at greater frequency than females; substantial proportions of males may breed annually (Hamann et al., 2003). Newer techniques, such as ultrasound, are useful and minimally invasive for evaluating the reproductive condition of adult sea turtles of both sexes. If the kidney is located first as a landmark in the male, the size and density of the testis and epididymis (parts of the male reproductive system) can be determined and the diameters of epididymal tubules measured for comparative studies (Blanvillain et al., 2008). Male breeding rates will inform our understanding of the proportion of males in a population required for successful reproduction and our understanding of possible depensation effects. (Depensation is described in the section "Density Dependence" later in this chapter.)

Recruitment of females into the breeding population and the proportion of first-time breeders in a nesting population are critical for assessing population trends. For example, if a nesting population is increasing in abundance, is the increase the result of increased recruitment of first-time breeders, increased survival of mature females, or both? In nesting populations subject to saturation tagging (tagging of every female) for a duration longer than the remigration interval with no loss of individual identification through tag loss and no immigration due to low fidelity, recruitment can be measured directly as the number of females that arrive with no tags (Richardson et al., 2006; Dutton et al., 2007). Few studies, however, meet those requirements. Another technique, laparoscopy, can be performed on female sea turtles at rookeries to determine the proportion of females that are first-time breeders or performed on foraging grounds to assess the proportion of female recruits that are preparing to breed in that year (Hamann et al., 2003). However, a method that is less invasive and more rapid is needed to distinguish recruits from females that have nested in previous seasons.

FECUNDITY

Fecundity is the reproductive output of an individual or a population. In sea turtles, fecundity is usually measured as the number of eggs deposited during a nesting season, which when combined with breeding rate (see above) yields an estimate of lifetime fecundity (average breeding rate multiplied by average reproductive lifespan). Within a nesting season, egg output of an individual is the product of the number of clutches deposited (clutch frequency) and the number of eggs in each clutch (clutch size). Egg size is usually not considered a measure of fecundity. However, because egg size is both a predictor of offspring quality and a component of estimates of resource allocation to offspring production, it is included in this discussion. Production of healthy hatchlings is another and perhaps better measure of fecundity than the production of eggs; therefore, the committee also addresses fertility, temperature-dependent sex determination, and hatching success.

Egg Production

Clutch frequency is an extremely important demographic parameter for both population models and assessment of trends in population abundance. Many monitoring programs on nesting beaches rely on nest counts to generate estimates of and trends in population abundance with the explicit assumption that clutch frequency is constant, but clutch frequency requires continual monitoring because it may vary among years (Broderick et al., 2003). For example, clutch frequency varied substantially with quality of nutrition in green turtles (*Chelonia mydas*; Bjorndal, 1985), and this indicates that changing resource and environmental conditions affect clutch frequency. Attempts to measure clutch frequency have been based largely on saturation-tagging programs on nesting beaches. Because of the length of the reproductive interval and the distance over which females deposit nests in a given season, intercepting females at each emergence is challenging. In Florida, for example, individual loggerheads (*Caretta caretta*) have been recorded nesting up to eight times in one season over an 82-day interval (Tucker, 2009), and one female deposited nests over a range of 182 km along the east coast of Florida during one season (Bjorndal et al., 1983). Hence, many published estimates of clutch frequency need to be viewed with caution.

Other approaches have been used to estimate clutch frequency and deserve further development. Radio and satellite telemetry have both been used. Radio telemetry is limited by the relatively short transmission distance and labor-intensive nature of monitoring. The relatively large location error of satellite telemetry has limited its application but does not preclude its application (Tucker, 2009). This technology will become more valuable as telemetry systems that generate more accurate locations are developed. Rivalan et al. (2006b) estimated clutch frequency in leatherbacks in French Guiana by using mark–recapture data to model stopover duration. A recent initiative used genetic markers from one egg in each clutch deposited in Georgia to identify the individual female that had deposited the clutch and thus the number of clutches deposited by each female (Brian Shamblin, personal communication). Methods of estimating clutch frequency that are relatively inexpensive and can be applied repeatedly on nesting beaches around the world are greatly needed.

Clutch size may be the only demographic parameter on which there are adequate data. The most accurate counts of clutch size are made during the egg-laying process, but with proper training and experience, accurate egg counts can be determined from pieces of egg shells during nest inventories after hatchlings have emerged (Miller, 1999). Unlike clutch frequency, clutch size apparently is not greatly affected by environmental factors (Bjorndal, 1985; Bjorndal and Carr, 1989) although it does vary substantially within populations or individuals over time (van Buskirk and Crowder, 1994; Broderick et al., 2003). Female body size accounts for some of the variation, as does time within the nesting season (Frazer and Richardson, 1985; van Buskirk and Crowder, 1994; Broderick et al., 2003). A better understanding of the reasons for the variation would be valuable for determining the importance of clutch size as a basis of population assessment.

For a parameter that is so easily measured, there are surprisingly few data on sea-turtle egg size. Egg size is measured most commonly as egg diameter, but egg mass and volume have also been measured. Egg size varies widely among sea-turtle species (van Buskirk and Crowder, 1994) and perhaps between populations and individuals of the same species. Substantial variation in hatchling size has been shown recently in flatback (*Natator depressus*) populations (Whiting et al., 2008). Accounting for the variation in egg size, evaluating the relationship of egg size to hatchling size (see van Buskirk and Crowder, 1994), and determining whether egg size is substantially affected by environmental factors would be valuable in assessing the importance of egg size as a factor in population assessment.

Hatchling Production

Survival from egg deposition to the emergence of hatchlings from the nest is the best quantified life stage of sea turtles. Given the accessibility of this stage, however, the number of quantitative studies, particularly for natural nests that have been subject to management interventions, is surprisingly low (National Research Council, 1990). The paucity of published data is due primarily to three factors: the difficulty of marking and following nests, the substantially longer monitoring period that is required to quantify hatching success throughout the season, and the lack of publication of many of the studies.

Determining hatching success is critical for the assessment of sea-turtle populations. Relying solely on the number of nests deposited to estimate hatchling production can lead to serious overestimates. In Tortuguero, Costa Rica, which has the largest green turtle Atlantic nesting population, Horikoshi (1992) reported that hatchling success was substantially reduced by high groundwater that drowned many nests, although no problem was apparent in surface observation of the beach.

Many natural and anthropogenic factors can affect embryo survival and reduce hatching success (Lutcavage et al., 1997). Techniques for evaluating hatching success have been summarized by Miller (1999). Loss to predators—both natural predator populations and those introduced or subsidized by humans—can be high (Stancyk, 1982). For example, raccoon populations that had increased above natural numbers as a result of human activities were responsible for predation of up to 97% of loggerhead nests on some Florida beaches (National Research Council, 1990).

Although the fertility of eggs deposited by sea turtles is generally high, probably exceeding 95% (Miller, 1997; Bell et al., 2003), low egg fertility can be a problem so egg fertility needs to be monitored in studies of hatching success. Decreased egg fertility of leatherback eggs in Terengganu, Malaysia (Chan, 1989), probably resulting from a reduction in the ratio of males to females, has been identified as a factor in the dramatic decline of nesting in that rookery (Chan and Liew, 1996).

All species of sea turtle exhibit temperature-dependent sex determination (Wibbels, 2003). That is, the temperature at which an embryo develops is primarily responsible for determining the sex of the hatchling (but see LeBlanc and Wibbels, 2009). In sea turtles, females are produced at higher temperatures and males at lower temperatures. Therefore, the primary sex ratio—the sex ratio of hatchlings—can vary greatly among clutches, among months within a nesting season, among nesting seasons, and among nesting beaches. Environmental changes, such as construction of tall buildings in Florida that shade the beach and reduce sand temperatures (Mrosovsky et al., 1995) and removal of trees behind the nesting beach in Terengganu that result in higher sand temperatures (Chan and Liew, 1996), can have substantial effects on the sex ratio of hatchlings. Hatchling sex can be identified reliably only with gonad histology or morphology (Ceriani and Wyneken, 2008); a nonlethal, accurate technique that could be used on a large number of hatchlings is greatly needed (Wibbels, 2003). Such a technique will be critical for monitoring responses of populations to climate change. As temperatures increase, primary sex ratios may shift toward females. Because many nesting beaches already produce primary sex ratios strongly biased toward females, there is concern that the proportion of males will be insufficient and that fertility of eggs could decline (Hawkes et al., 2007; Poloczanska et al., 2009). Laparoscopy may constitute a nonlethal technique for determining sex in hatchlings (Wyneken et al., 2007) and needs to be investigated.

SURVIVAL PROBABILITIES

One of the greatest gaps in developing the conceptual model is in estimates of survival of immature turtles and nesting females of all species. Survival of turtles through embryonic development to their emergence from the nests is discussed above (see the section "Fecundity").

Adult Females

Estimates of survival for adult females have been derived from mark–recapture studies that used open robust design for hawksbills (Kendall and Bjorkland, 2001) and leatherbacks (Dutton et al., 2005). This analysis is the best available approach for estimating survival probabilities based on mark–recapture data on nesting beaches if sufficient data are available. Survival estimates have also been generated from recovery analyses (Campbell and Lagueux, 2005; Troëng and Chaloupka, 2007) and a model of remigration intervals (Solow et al., 2002). Applying more than one approach to a population can increase confidence if the independently derived estimates are similar. For green turtles nesting at Tortuguero, Costa Rica, four analyses that used three techniques yielded similar estimates of probabilities of annual survival of adult females (Solow et al., 2002; Campbell and Lagueux, 2005; Troëng and Chaloupka, 2007).

Despite multiple calls for new studies (see Table 1.2; Turtle Expert Working Group, 2000; Heppell et al., 2003), there have been few attempts to update estimates of survival of loggerhead turtles nesting in the United States with mark–recapture analysis (e.g., Hedges, 2007), and current models still rely on results from the 1970s when mark–recapture studies were conducted on Little Cumberland Island, Georgia (Richardson et al., 1978; Frazer, 1983). The survival rates from those studies were not estimated with the open robust-design methods that have been developed to account for detectability of nesting females (Kendall and Nichols, 2002) but did account for tag loss. Efforts to assess loggerhead status and interpret trends in nests with lifecycle and simulation models have been stymied by the lack of new estimates (Turtle Expert Working Group, 2000; National Marine Fisheries Service Southeast Fisheries Science Center, 2001). That has also prevented proper evaluation of the effectiveness of management actions, such as the implementation of turtle excluder devices (Epperly and Teas, 2002).

Survival of nesting female Kemp's ridley (*Lepidochelys kempii*) turtles was estimated in a model-fitting exercise in which a simple age-structured model was fitted to nest census counts from Mexico to obtain a point estimate of annual survival before and after 1990 (Turtle Expert Working Group, 2000; Heppell et al., 2005). That was a unique circumstance in that all nesting of this highly endangered species was restricted largely to one well-monitored nesting beach, and the population had exhibited changes in trends that provided contrast for model fitting. However, the estimate was not empirically based, and additional analysis of existing mark–recapture data on females tagged with passive integrated transponders (PITs) is needed (Heppell et al., 2007).

Immature Turtles

The paucity of estimates of annual survival of immature sea turtles on their oceanic and neritic (nearshore) foraging grounds limits the ability to assess sea-turtle populations. Mark–recapture models based on tagging studies (Chaloupka and Limpus, 2002, 2005; Bjorndal et al., 2003c; Campbell and Lagueux, 2005; Braun-McNeill et al., 2007) and catch-curve analyses (Frazer, 1987; Bjorndal et al., 2003b) have been used to generate estimates. A serious limitation of both approaches, particularly in Atlantic populations in which immature turtles tend to move among foraging grounds to a greater extent than in the Pacific, is the confounding of emigration and mortality in estimates of apparent survival (usually referred to as phi). Differences between apparent survival and true survival can be substantial in populations of immature sea turtles (Bjorndal et al., 2003c). Estimates of survival not confounded with emigration are possible with Burnham models (Burnham, 1993; Catchpole et al., 1998), joint analyses of live-recapture and dead-recovery data (Bjorndal et al., 2003c; Seminoff et al., 2003), if sufficient data are available. Transients, which are usually identified as marked animals seen only once in a study area, can lead to biased estimates of survival probability (Pradel et al., 1997). Accounting explicitly for transient behavior of marked sea turtles has been undertaken in a few studies of sea-turtle survival probabilities (Chaloupka and Limpus, 2002; Sasso et al., 2006) but needs to be explored further. Another common technique in fisheries, catch-curve analyses, requires knowledge of size-at-age, which can limit applications to sea-turtle populations, and needs to incorporate differential growth rates and recruitment.

Data on strandings of sea-turtle carcasses cannot be used to estimate survival probabilities. However, stranded carcasses can be used to assess abrupt changes in mortality due to changes in fisheries or disease outbreaks or to track the incidence of diseases. Stranding data are most valuable in hazard-specific analyses (Crowder et al., 1995; Chaloupka et al., 2008b) because the proportion of the population represented by stranded turtles is unknown.

A major anthropogenic hazard for sea turtles worldwide is incidental capture in shallow-set pelagic longline fisheries (Lewison et al., 2004). Many turtles caught in such fisheries are alive when released from the gear (Gilman et al., 2007), but it is widely assumed that a substantial number will die soon after because of injuries caused by hooks or line entanglement (Lewison et al., 2004). However, there are few reliable estimates of post-release mortality in sea-turtle species despite their being essential for risk assessment and hazard mitigation. Chaloupka et al. (2004a) and Sasso and Epperly (2007) used satellite telemetry to estimate post-hooking mortality in loggerhead sea turtles but pointed out limitations of the method, including inadequate sample sizes and premature release of satellite tags, that make it difficult to derive reliable cause-specific mortality estimates.

DISPERSAL PROBABILITIES

Movement of Adult Females Between Rookeries

To date, all measured probabilities of female movements between rookeries are too low to influence management plans. Nesting females are highly philopatric (i.e., they return to their birthplace), but the degree of site specificity varies among species. Loggerhead nesting populations may show population structure (mitochondrial DNA differentiation) on a scale of less than 100 km (Bowen et al., 2005), green turtles on a scale of 500 km (Dethmers et al., 2006), and olive ridley (*Lepidochelys olivacea*) and leatherback turtles on a scale of more than 500 km (Lopez-Castro and Rocha-Olivares, 2005; Dutton et al., 2007). That information is important because the degree of site specificity and the scale of population structure determine the appropriate sizes of management units (see Chapter 2) and determine the extent to which nesting populations will reinforce each other.

Those geographic scales are supported in some species by tag-recapture data from renesting females. However, long-distance relocations of nesting females (beyond the geographic ranges outlined above) have been documented. LeBuff (1974) reported a loggerhead female relocating from southwest Florida to southeast Florida, and at least two tagged females have switched from Tortuguero to other locations in the Caribbean (citations in Bowen et al., 1992). A low level of switching between nesting sites is beneficial and probably necessary for the long-term persistence of sea-turtle species. In view of epochal changes in climate, oceanography, and geography, the appropriate nesting sites of, for example, the Pliocene are not the same as the ones today. Shifting among nesting beaches allows sea turtles to respond to a changing world.

Dispersal of Immature Sea Turtles

Immature sea turtles generally undergo two phases of dispersal (both of which are poorly understood): (1) hatchlings disperse away from the nesting beach into oceanic habitats after emergence from the nest and (2) immature turtles disperse from oceanic habitats when they recruit to neritic habitats, usually years before reaching sexual maturity. Once on neritic foraging grounds, immature turtles tend to move among foraging habitats. Knowledge of movements of immature sea turtles has improved through increased flipper tagging of immatures, satellite telemetry, genetics, and stable isotopes and has revealed a more complex series of dispersals of some turtles (Eckert and Martins, 1989; Eckert, 2002; Bolten, 2003a; Harrison and Bjorndal, 2006; McClellan and Read, 2007; Reich et al., 2007).

Evaluation of dispersal of hatchlings has been limited to direct observations (Frick, 1976; Witherington, 1991), tissue transplants or "living tags" (Wood and Wood, 1985), shell notching (Limpus, 2009), and evaluation of current patterns (Blumenthal et al., 2009b). Over 43,000 Kemp's ridley hatchlings were marked with internal wire tags in 1996–2000 (Caillouet, 1998; Snover et al., 2007). All those techniques have well-documented limitations. The greatest challenge for any mass-hatchling tagging program (e.g., with wire tags or PIT tags) is to intercept and recognize these marked turtles in their juvenile stages. The feasibility of an improved program of marking large numbers of hatchlings so that they can be recognized when they appear in oceanic or neritic foraging grounds could be explored.

In 2009, neonate loggerheads were tracked successfully with highly miniaturized satellite transmitters that had been designed for birds (Mansfield et al., 2009b). In addition, application of hatchling-dispersal models coupled with multitrophic biophysical models,[1] such as the Spatial Ecosystem and Population Dynamics Model (Lehodey et al., 2008), now being applied to pelagic fish can be used to predict movements and habitat occupancy through the first years of life.

The recruitment of sea turtles from oceanic to neritic habitats can occur over a range of sizes and, presumably, ages (Bolten, 2003b). Sufficient numbers of recruits from the mass tagging of Kemp's ridley hatchlings with internal wire tags were identified to estimate the age of recruitment as 2.2 years (Dodge et al., 2007). Identifying new recruits on neritic foraging grounds is challenging; a number of techniques have been used but with uncertain success. Arrival of turtles without tags in areas with saturation tagging (Bjorndal and Bolten, 2008) with epibionts (organisms that live on the surface of other living organisms) from oceanic habitats (Limpus and Limpus, 2003a) or, in green turtles, with clear plasma color (Bolten and Bjorndal, 1992) has been used to identify recruits. Stable iso-

[1] Models that integrate effects of biological and physical parameters over several trophic levels.

tope signatures of carbon and nitrogen in scute tissue (the keratin covering of the upper shell that is inert after deposition) provide a history of diet and habitat that can be used to identify recent recruits (Reich et al., 2007). Reliable, rapid, and non-invasive methods of identifying recruits are needed.

SOMATIC GROWTH AND AGE AT SEXUAL MATURITY

Somatic growth has been measured in a number of sea-turtle populations. Adult females essentially stop growing after attaining sexual maturity, at which point resources are allocated away from somatic growth to reproduction. In immature turtles of a given species, growth varies spatially and temporally (Diez and van Dam, 2002; Balazs and Chaloupka, 2004b; Chaloupka et al., 2004b; Kubis et al., 2009). Known sources of variation are body size (Chaloupka and Musick, 1997), population density (Bjorndal et al., 2000a), habitat quality (Diez and Van Dam, 2002), nutrient quality of diet (Wood and Wood, 1981), disease status (Chaloupka and Balazs, 2005), and compensatory growth (Bjorndal et al., 2003a; Roark et al., 2009a). A combination of somatic growth rates with indexes of body condition is the best current measure of habitat quality and population status on foraging grounds (Bjorndal et al., 2000a; Diez and van Dam, 2002; Kubis et al., 2009).

The most common method of measuring growth rates in turtles has been mark–recapture study. Because population and environmental conditions can be monitored throughout a mark–recapture study, this technique offers the best approach for evaluating the mechanisms that regulate growth. Mark–recapture studies are of necessity long term and labor intensive and are successful only when recapture probabilities are relatively high. Because that condition is not always met, other techniques have been used.

Skeletochronology, the use of markers in skeletal material (primarily humeri and eye ossicles), has been used in many studies to estimate somatic growth rates (Zug et al., 1986; Bjorndal et al., 2003a; Snover and Hohn, 2004; Snover et al., in press). Caution in the interpretation of marks is critical, the technique is not practical for live animals, and remodeling of internal bone layers can be problematic. Those and other challenges in the application of skeletochronology have been well reviewed (Snover et al., 2007; Avens et al., 2009). Advantages of the technique are that turtles do not have to be captured, skeletal elements can be gathered from the large number of carcasses that strand on the U.S. coast each year, and longitudinal sampling of individuals can be exploited. Longitudinal sampling is possible only with multiple recaptures in mark–recapture studies. With skeletochronology, growth-increment analysis of the humeri can be used to detect individual variance in growth rates (Vaughan, 2009). A greater effort needs to be made to archive humeri from sea-turtle carcasses of known size, sex, location, and date for age and growth studies.

Length-frequency analyses, which rely on maximum-likelihood algorithms to detect ageclass modes in size distributions, have been used widely in fisheries and successfully in sea turtles (Bjorndal et al., 1995, 2000b, 2001). A disadvantage of the technique is that, with currently available software, only von Bertalanffy growth models can be used. Greater overlap of body lengths in older ageclasses may limit the use of the technique. Its main advantage is that it requires only data on size distributions.

Two other techniques for measuring growth of sea turtles have been investigated. Hays and Marsh (1997) estimated growth rates of the very early stages by analyzing drift times to remote locations and the size of small turtles at those locations. And the use of RNA and DNA ratios, which have been used extensively in studies of fish growth, has been tested in sea turtles with some success (Roark et al., 2009b). Both techniques deserve further evaluation.

Age at sexual maturity is a critical demographic parameter. Estimating age at maturity on the basis of somatic growth rates is problematic because, for all Atlantic populations, few data are available on growth rates of large subadult turtles (i.e., above 70 cm in carapace [upper shell] length in green turtles). A high priority might be to determine growth rates of large subadults so that estimates of age at sexual maturity can be based on a stronger foundation.

SEX RATIOS

Because sea turtles exhibit environmental sex determination, primary sex ratios are determined by environmental factors, as described above (see the section "Fecundity"). Variation in secondary sex ratios (i.e., the odds of a hatchling will be male) on foraging grounds may result from variation in primary sex ratios, sex-specific mortality, or sex-specific dispersal. Data on secondary sex ratios of immature and adult sea turtles are needed to develop sex-specific population models and to evaluate "optimal" sex ratios (i.e., ratios at which reproductive output is maximized in a population). If it becomes necessary because of global warming, the latter will be critical for programs to manipulate primary sex ratios at nesting beaches (Mrosovsky and Godfrey, 1995).

DENSITY DEPENDENCE

Rates in a population are said to be density dependent if they vary with the abundance or density of the population. For example, in the classic logistic model of population growth, the per capita population growth rate increases linearly as the population declines. That kind of density dependence is termed compensatory because it tends to stabilize population size. When the population is small, the per capita growth rate is high, and the population increases toward its carrying capacity. As the population nears its carrying capacity, the growth rate declines as births and deaths become equal, and the population reaches a stable abundance. If other characteristics are equal, a depleted population will begin to recover relatively rapidly when compensation is present if the limiting factor (e.g., harvesting, bycatch, disease) is reduced or eliminated. The most common cause of compensation is competition for food, space, or other resources.

In contrast with compensation, some populations exhibit a form of density dependence termed depensation, in which over some low range of abundance the per capita growth rate decreases as abundance declines. Depensation is said to be critical when the per capita growth rate becomes negative at low abundance. When depensation is operating, a depleted population will tend to recover slowly and therefore be vulnerable to extinction shocks. If depensation is critical, the population may become extinct despite the elimination of the limiting factor. Although the classic explanation of depensation (also referred to as the Allee effect) is the rarity of high-quality mating opportunities, other factors may be involved (Liermann and Hilborn, 2001).

Turtle population growth has been evaluated with stage-based matrix models that typically assume that vital rates are independent of density. That reflects primarily a lack of information about density effects on these rates. The effect on model predictions of ignoring density dependence remains an open question. Clearly, if vital rates are strongly density dependent, a model with fixed rates will be, at best, applicable over a narrow range of population size. Despite that limitation, model predictions may be correct qualitatively (Heppell et al., 2000). Chaloupka and Balazs (2007) developed a statistical state-space model for Hawaiian green turtles that allows for density dependence (either compensatory or depensatory).

Of necessity, most work on identifying density dependence in sea-turtle populations has focused on processes that occur on nesting beaches. For example, Girondot et al. (2002) identified density-dependent nest destruction in the leatherback turtle on a beach in French Guiana; the rate of destruction increased with nesting numbers (presumably as a result of nesting-habitat limitation). That was accompanied by a density-dependent feminization of the hatchling sex ratio (Caut et al., 2006). Tiwari et al. (2006) also reported density-dependent nest destruction and predation on hatchlings in a Caribbean green turtle population. In a study not involving nesting processes, Bjorndal et al. (2000a) found evidence of density dependence in the somatic growth rate of immature green turtles in the Caribbean region. They found a negative correlation between population density and both the mean annual growth rate (as measured by carapace length) and an index of body condition. That suggests that Caribbean green turtles are food limited when abundance is high.

Bell et al. (2010) evaluated evidence of depensation in green turtles and loggerhead turtles. They focused on the relationship between rookery size (as measured by total clutches per season) and fertilization success, hatch success, and hatchling emergence success; they used data on the Cayman Islands and a meta-analysis of global data. The study found no evidence of depensation in either species in either the Cayman Islands data or the global data. However, because the analysis was based on a mixture of cross-sectional and time-series data, the result needs to be treated with caution. A more complete analysis would treat the data as multiple time series with depensation operating within, but not between, the component series.

STRANDINGS DATA

A substantial proportion of the effort expended to collect sea-turtle data in the United States is invested in the Sea Turtle Stranding and Salvage Network (STSSN). Because the usefulness of the data generated in that program has been debated (Epperly et al., 1996; Turtle Expert Working Group, 2000), the committee addresses STSSN here in some detail. Sea-turtle strandings occur when animals have washed up on a beach or into shallow water. Stranded animals may be dead or dying because of anthropogenic causes, such as interactions with fisheries, or natural morbidity, such as disease or "cold stunning" when they have been exposed to lethal cold-water temperatures. Strandings include all life stages that are present in neritic habitats, including juveniles and adult males. Carcasses provide opportunities for data collection that are difficult or impossible with live animals, such as collection of data for evaluating maturational status and removal of the humeri for age and growth studies. Carcasses are checked for tags; this is an important source of tag recoveries that are used to evaluate growth and dispersal of individual turtles. Strandings can also provide some information on mortality and have been correlated with levels of fishing effort and enforcement (Lewison et al., 2003). With careful consideration of the many sources of variability that affect the probability of stranding and detection of carcasses, strandings may also provide distribution and trend information that is relevant to population assessment (Chaloupka et al., 2008b).

The density of strandings has been used as a trigger for management action in some areas and has resulted in spatial closures in fisheries

(Santora, 2003). Strandings have been clearly linked to fishing activity (Caillouet et al., 1996; Chaloupka et al., 2008b), and changes in the relative abundance of strandings have been used as an indicator of management effectiveness (Crowder et al., 1995; Lewison et al., 2003). However, many factors are related to the frequency of strandings, including cause of death and condition of the carcass, location of death and water currents, water temperature (which affects the decay rate), and salvage effort. Physical processes that affect stranding rates may also change, and this potentially necessitates regular evaluation of the correlation between ocean conditions and the probability of stranding as climate-driven forcers vary in time and space. Concerns about these caveats have led to disagreement about the value of strandings for population assessment (Epperly et al., 1996; Turtle Expert Working Group, 2000). Expert working groups and recovery-planning teams have agreed that strandings are highly stochastic events that provide information about local mortality events and a minimum estimate of regional mortality, but it may be difficult to extrapolate trends in strandings to changes in population abundance. Strandings account for an unknown proportion of total mortality and probably varies among regions. Nevertheless, patterns of strandings in time and space can provide information about seasonal distribution and interactions with fisheries when carcass-recovery efforts are standardized and data are pooled over broad spatiotemporal scales (Chaloupka et al., 2008b; Tomás et al., 2008).

STSSN in the United States operates in each coastal state and is coordinated through the National Marine Fisheries Service (NMFS) and state agencies. The network is run by a state coordinator and depends heavily on local volunteers. Coordinators are responsible for training programs for the volunteers and for the weekly or bi-weekly data reports that are sent to NMFS. Although the specific goals of each salvage program vary, most are designed to evaluate carcass abundance and trends that are assumed to be indicative of the living population of turtles in the monitored area. The programs provide data that can be used to quantify seasonality, species composition, population structure, life-history stage, sex ratio, and spatial distribution of turtles that wash ashore.

All STSSN-recovered animals are identified as to species, checked for external tags, and recorded by date and location. Carapace length and general condition of the carcass are also recorded for most animals. Many recovered dead turtles are necropsied by the state coordinator and staff to identify sex and state of maturation, to record ingestion of plastic (Bjorndal et al., 1994), and to conduct a general evaluation of the potential cause of death (although this can only rarely be determined). Carcasses may also be checked for PIT tags or magnetic wire tags. Some samples are collected from necropsied animals for specific projects, including tissue samples for contaminants evaluation and bones (humerus and eye ossicle) for aging (Snover and Hohn, 2004), and body-size data have been used to generate curves of somatic growth (Bjorndal et al., 2001). The proportion of recovered turtles that are evaluated thoroughly varies by state and frequency of strandings. Although sampling of sea-turtle carcasses for specific research projects does occur, the extent to which samples are collected and archived is variable in the southeastern United States, where most sea-turtle strandings occur (Laris Avens, personal communication). In Hawaii, the Pacific Islands Fisheries Science Center regularly archives tissue and humeri samples from recovered turtles. Strandings data are compiled and reviewed by the Southeast Fisheries Science Center, but researchers must request access to the data from each state individually, and state coordinators vary in their criteria for sharing data.

Changes in size distributions of strandings may be a valuable indicator of shifts in age structure or distribution of juveniles (Shoop et al., 1999). Turtle Expert Working Group reports (2000, 2009) have included examinations of trends in total strandings by region and the size distributions of strandings. Kemp's ridley and loggerhead strandings size distributions were converted to age distributions with an age-length key and used to estimate total instantaneous mortality with a simple catch-curve method (Turtle Expert Working Group, 2000). This approach raises a number of issues that were discussed by the report authors, including the unknown relationship between sizes of strandings and those of the population-at-large, the need to pool strandings across several years because of small samples, and variable growth rates that confound the age-length relationship. Recently, a data review by a loggerhead turtle working group included plots of turtle sizes observed through time that showed a good correlation between the size distributions of strandings on the east coast of the United States and of turtles observed at a power-plant intake in Florida and juvenile mark–recapture surveys (Turtle Expert Working Group, 2009; Vaughan, 2009). That suggests that strandings may be a reasonable indicator of what turtles are in the nearshore population, at least on broad spatial and temporal scales. Confirmation of the congruence is needed, particularly if researchers want to continue to use strandings to estimate mortality.

If the size composition of the nesting population is known and a strandings event can be linked to a particular fishery or environmental event, the observed size distribution can help to determine selectivity of the mortality source (what size classes are susceptible to the fishing gear used or an environmental event, such as red tide). Further research on the "selectivity curve" for strandings would be helpful to determine how the probability of carcass recovery is affected by the size and condition of the animal and its environment. A study using drifter bottles deployed in the South Atlantic Bight provided a rough estimate of a 20% probability of reaching shore for a wind- and current-driven carcass, with strong seasonal and spatial variability (Hart et al., 2006); similar studies need to be conducted in the Gulf of Mexico and the northeast United States, with an emphasis on establishing the likelihood of detection and statistical discrimination among spatial scales (Wiens, 1989).

The level of environmental monitoring needed to identify relationships between oceanography and strandings may be substantial, given the complexity and variability of nearshore ocean processes. It is possible that environmental variance will nullify strandings as a source of trend and distribution information for nesting populations, but large strandings events are still valuable indicators of local conditions (e.g., harmful algal blooms, intense fishing mortality), and samples from dead animals can provide important information through diet evaluation about local population structure, growth rates, maturation rates, and habitat use.

Every recovered carcass can be a valuable source of information for assessment if recovery efforts are standardized; proper measurements are taken; and samples are collected, processed, and archived according to established protocols. To improve the value of strandings data for assessment, each state program needs to be reviewed and evaluated for consistency in recovery effort, volunteer training, and protocols. Areas that have low or inconsistent sampling effort could be identified to improve extrapolation methods. Programs for evaluating size distributions and growth rates from turtle hard parts need to be supported and enhanced to maximize the amount of information obtained from each stranded animal. Flipper collection could become standard protocol for STSSN volunteers in the southeastern and Gulf states, but a considerable investment in time and resources will be needed to process and evaluate those samples.

RECOMMENDATIONS

- Researchers should give high priority to generating estimates for the following parameters: survival of immature turtles and nesting females, age at sexual maturity, breeding rates, and clutch frequency.
- Because demographic rates can vary over time and space, researchers should collect data over both dimensions so that population trends can be detected and evaluated adequately.
- Researchers should be aware that evaluation of point estimates of demographic parameters is not sufficient for population assessment; characterizing uncertainty and variance is also necessary.
- Researchers should strive to understand the mechanisms regulating variation in demographic rates; this is essential for diagnosing changes in population abundance and mitigating population declines.

• NMFS and the U.S. Fish and Wildlife Service should arrange for a review of data now being collected under the auspices of, or with the support of, their agencies and evaluate the costs and benefits. For example, the sea-turtle stranding and salvage networks should be evaluated, perhaps with the assistance of the U.S. Geological Survey's National Wildlife Health Center.

• STSSN should collect—in addition to data on abundance, size, condition, and sex—samples of tissues and hard parts that can be used to identify stock of origin, to assess diet through isotope analysis, and to evaluate age and growth.

6

Integrating Demographic Information with Abundance Estimates

Sea-turtle management has been focused on reducing mortality from as many sources as possible on all possible life stages. That is a laudable goal for any endangered species, and it is reasonable to assume that minimization of anthropogenic mortality would result in population recovery. Yet, in spite of decades of monitoring and litigation, some U.S. populations (e.g., northwest Atlantic loggerheads [*Caretta caretta*]) do not appear to be recovering, and the status of most is unknown or inferred exclusively from nesting-beach trends (see Table 1.1).

Wildlife and conservation researchers understand that using abundance measures of a single life-history stage can be misleading in diagnosing the status and trends of a population (Van Horne, 1983; Thomson et al., 1997; Brooks et al., 2004), including the diagnosis of sea-turtle trends (Bjorndal et al., 1999; Hays, 2000; Chaloupka, 2001b; Solow, 2001; Chaloupka and Limpus, 2002; Heppell et al., 2003). Integrating abundance measures with demographic processes in a framework of modeling and data fitting provides a more robust basis for diagnosing trends, evaluating the effects of anthropogenic hazards, and defining recovery criteria (Brooks et al., 2008).

In this chapter, the committee reviews some of the quantitative tools used in assessment of populations, reviews which tools have been applied to sea-turtle assessments and discusses the procedures that are routinely used in fishery assessments to ensure scientific rigor and could be adopted for future assessments of sea turtles.

MODELS FOR POPULATION ASSESSMENT

Mathematical models are powerful tools for species assessment and evaluation. The reliability and utility of models depend on the quality and availability of data and on the assumptions conferred by model structure. Population models for sea turtles have been reviewed by Chaloupka and Musick (1997), Heppell et al. (2002), and others. Published models have ranged from regression fits to nesting-numbers data, deterministic lifecycle analyses, and complex simulation models—all with varied data requirements and assumptions. There are tradeoffs in model construction among precision, realism, and generality. Levins (1966) argued that a particular model can achieve at most two of those three qualities. Appropriate model complexity depends heavily on the question asked. The results of a simple model might be robust in uncertainty in lifecycle parameters but qualitative or incapable of supporting the precise estimates of population size or the effects of removal of individuals from a population. In contrast, detailed simulation models may require a large amount of biological information to produce precise or reliable estimates of population size or to predict response to perturbations. Regardless, models that are to be used for assessment, prediction, and management decisions require solid demographic data, preferably as time series of information that can be analyzed for changes in response to stressors, population density, or environmental variability (Hilborn and Mangel, 1997).

TOOLS FOR ASSESSMENT

Sea-turtle management issues vary by region, but quantitative assessment generally focuses on the following four primary issues:

- Evaluation of trends in nesting and foraging population abundance as an indicator of population status
- Diagnosis of the potential causes of those trends
- Evaluation of the effects of natural and anthropogenic hazards on population viability
- Definition of recovery criteria

Here the committee reviews a variety of available modeling approaches to questions about sea-turtle status and trends and notes the data requirements for each (Table 6.1). Unlike fishery assessment, the focus for sea-turtle management in the United States is not on sustainable harvest. Nevertheless, many of the quantitative tools used in fishery assessments are applicable to sea turtles and other threatened species. The results of the approaches identified here vary from qualitative to highly quantitative (Table 6.1).

TREND EVALUATION AND EXTINCTION RISK

Trends in Abundance and Abundance Indexes

The most common evaluations of sea-turtle population status are those of nesting-beach trends, which may be based on counts of nests or nesting females (see Chapter 4). Linear regression is often used to identify an exponential growth rate for each nesting beach and used with data that is pooled by region (e.g., National Marine Fisheries Service Southeast Fisheries Science Center, 2001). Regression methods have also been used to evaluate trends in abundance indexes derived from juvenile and adult sampling at sea. Slopes and confidence intervals from simple regression analysis are easy to interpret but may fail to include important biological complexities that relate what is counted (such as nests) to a trend at the population level. The numbers of nests or nesting females may be highly variable because of environmental effects on the probability of breeding and other factors (Solow et al., 2002) so data are sometimes smoothed by using a running sum or averaging (e.g., Turtle Expert Working Group, 2007; Snover and Heppell, 2009).

Uncertainty in population trends has been evaluated with Bayesian state-space methods that are not restricted to parametric statistical evaluation and permit a more transparent evaluation of the probability of population decline (Turtle Expert Working Group, 2007, 2009). In the Bayesian approach, trends are expressed as probabilities of increase or decline rather than as slopes and confidence intervals but still require biological information for extrapolation of nest counts to population abundance. More complex trend-evaluation models that incorporate environmental drivers, such as nonparametric regression or Bayesian generalized additive models (Bjorndal et al., 1999; Chaloupka, 2001b; Balazs and Chaloupka, 2004b; Troëng and Rankin, 2005), have also been applied. The advantage of the Bayesian approach is that the confidence intervals do not require normal approximation assumptions but are based on the data themselves, and this provides a natural means of evaluating both sampling uncertainty and process error caused by environmental variance.

Without estimates of breeding probability (remigration interval) and of recruitment of new turtles to the breeding population, assessment of population trends on the basis of nesting-beach data is highly tenuous. A change in the number of nests may be due to a change in the frequency of nesting, a change in adult-female survival, or a change in the number of first-time breeders, none of which is monitored by the agencies. Estimates of trends in juvenile-turtle abundance through in-water surveys, aerial surveys, and frequency of strandings have generally been evaluated with regression analysis after an evaluation of data uncertainty (e.g., Turtle Expert Working Group, 2009).

TABLE 6.1 Common Evaluation Methods and Modeling Tools That Have Been Applied to Sea-Turtle Assessment and Their Basic Data Requirements.[a]

Focus	Method	Accuracy	Quantitative	Abundance
Trend evaluation	Linear regression of abundance index (nests)		Yes	X
	Bayesian trend evaluation		Yes	X
	In-water trends		Yes	X
	Diffusion approximation		Yes	X
Trend diagnosis	Surplus production	lower	Yes	X
	Transition matrix	↓	Yes	
	Aggregate simulation	↓	Yes	X
	Individual-based simulation	↓	Yes	
	Integrated models	↓	Yes	X
	Ecosystem models	higher	Yes	X
Evaluating anthropogenic impacts	Bayesian belief network	lower	No	
	Diffusion approximation	↓	Yes	X
	Potential biological removal	↓	Yes	X
	Surplus production	↓	Yes	X
	Aggregate simulation	↓	Yes	X
	Individual-based simulation	↓	Yes	X
	Integrated models	↓	Yes	X
	Ecosystem models	higher	Yes	X
Defining recovery criteria	Diffusion approximation	lower	Yes	X
	Aggregate simulation	↓	Yes	X
	Individual-based simulation	↓	Yes	X
	Integrated models	↓	Yes	X
	Ecosystem models	higher	Yes	X

[a] Methods are grouped according to three primary needs for management and ordered along a general gradient from lower to higher accuracy of model output. Increased accuracy is tied to model complexity and the need for detailed biological information.

Stochastic Projections and Diffusion Approximation of Extinction Risk

The simplest form of population-viability analysis projects a time series of abundance or an index of abundance and evaluates the probability of extinction (or recovery) on the basis of the proportion of projections that cross a predetermined threshold (Dennis et al., 1991; Holmes, 2001, 2004; Snover and Heppell, 2009). The model relies on estimates of the exponential trend and variance estimated from census data and can

Vital Rate						
Breeding Frequency	Clutch Frequency	Adult Survival	Juvenile Survival	Age at Maturation	Dispersal	Trophic Dynamics
X	X					
X	X				X	
X	X					
X		X				
X	X	X	X	X		
X	X	X	X	X		
X	X	X	X	X	X	
X	X	X	X	X	X	
X	X	X	X	X	X	X
X	X					
		X		X		
X						
X	X	X	X	X		
X	X	X	X	X	X	
X	X	X	X	X	X	
X	X	X	X	X	X	X
X	X					
X	X	X	X	X		
X	X	X	X	X	X	
X	X	X	X	X	X	
X	X	X	X	X	X	X

be evaluated analytically with a model that describes a diffusion process with drift, commonly referred to as a diffusion approximation of extinction risk (Dennis et al., 1991). Because time series of sea-turtle abundance are based on counts of nests or nesting females, the trend and variance through time must be adjusted to account for the relationship between nest number and adult-female number (clutch frequency, that is the number of clutches deposited by an individual turtle in a nesting season) and for autocorrelation (the similarity between observations as a function of the time separation between them) caused by remigration intervals (breeding frequency). The diffusion-approximation model has been applied recently to sea-turtle status assessment as a method of estimating trends and evaluating the risk of decline while accounting for uncertainty (susceptibility to quasi-extinction; Snover and Heppell, 2009). It has also been applied to evaluation of removals (Merrick and Haas, 2008; Snover, 2008), although the ability of the analysis to detect changes in extinction risk has yet to be evaluated fully.

Trends in an index of abundance and simple stochastic extinction risk can provide benchmarks for status determination, but it is the *diagnosis* of a trend that is more critical for decision making. Predicting how and why changes in abundance have occurred requires tools that provide additional biological details, particularly the mechanisms of population dynamics that are linked to the sea-turtle lifecycle (Chapter 3).

Surplus-Production Models

The surplus-production models is the most commonly used population-assessment approach when one is limited to datasets that consist only of harvest and relative abundance time series (Hilborn and Walters, 1992). Surplus-production models implicitly account for density-dependent demography—the change in population growth rate that is anticipated with changes in population size. The models do not include age structure but can be modified to include time lags. To determine parameter estimates through data fitting, the models require a time series of abundance data that can accurately demonstrate density-dependent population processes. Chaloupka and Balazs (2007) used a Bayesian state-space modeling approach to fit a stochastic surplus-production model to the Hawaiian green turtle (*Chelonia mydas*) nesting-abundance data series given the known commercial harvest history. This Bayesian-inference approach enabled prior knowledge of green turtle demography to be incorporated to supplement the limited information available on this population. The model accounted for both process and observation error. The approach also enabled uncertainty in model-parameter estimates and the temporal variability in nesting abundance to be accounted for explicitly. The main objective was to determine whether it was possible to derive useful estimates of population and management parameters for the Hawaiian green turtle population with the data available.

Age- and Stage-Structured Matrix Models

These structured models aggregate individuals into life-history stages or ageclasses, allowing incorporation of time lags. They can be deterministic or stochastic (random) and can (but often do not) include nonlinearities, such as density dependence (Caswell, 2001). Analytical sensitivity analyses of deterministic matrices have been used extensively for sea turtles to identify vital rates that have a large effect on asymptotic-model outputs, such as population growth rate and stage-specific reproductive value (reviewed in Heppell et al., 2003). Most deterministic matrix-model evaluations are useful for learning and discovery purposes to compare relative changes in abundance that may occur with changes in stage-specific vital rates (e.g., Crowder et al., 1994) or to compare qualitatively the potential effect of removals of turtles of different ages (Wallace et al., 2008). They can be used to predict population size only if vital-rate means and variances have remained relatively constant and if initial conditions of abundance and age structure can be determined. Matrix models that describe lifecycles can be simple or include complex population structure, such as the life stages shown in the conceptual model in Chapter 3; values are assigned to parameters on the basis of empirical estimates of survival, growth, or fecundity and estimates of dispersal if life stages are spatially explicit. Matrix models for simulation purposes can include assigning values to parameters through model fitting when time series of abundance, recruitment, or age structure are available (e.g., the model for Kemp's ridley [*Lepidochelys kempii*]; Turtle Expert Working Group, 2000; Heppell et al., 2005); however, this has not been done in most of the existing assessments because of uncertainty in age- or stage-specific vital rates and unknown population age structure. Diagnosis of observed population change can potentially be performed by using a life-table response experiment if the magnitude of the effects of different vital-rate changes in two or more periods can be evaluated (Caswell, 2001). Age-structured models used in fishery assessment, although not matrix models themselves, operate with the same principles of age-specific tracking through time and recruitment tied to adult abundance.

Stochastic Simulation Models

A number of stochastic, ageclass-specific, and individual-based simulation models have been developed to account for sea-turtle demography. Chaloupka (2003a) developed a stochastic simulation model for the southern Great Barrier Reef green sea-turtle population to foster better insight into regional metapopulation dynamics. The model (based on a system of ordinary differential equations) was sex-structure and ageclass-structure linked by various density-dependent, correlated, and time-varying demographic processes that are subject to environmental and demographic stochasticity. The density-dependent processes included depensatory or Allee effects that occur at low abundances when the per capita growth rate decreases as abundance declines. The simulation model was based on extensive demographic information derived for the population from a long-term sea-turtle research program established and maintained by the Queensland Parks and Wildlife Service. Model validation was based on comparison with empirical-reference behaviors, and sensitivity was evaluated by using multifactor perturbation experiments and Monte Carlo simulation within a fractional factorial sampling design. The model was designed to support evaluation of the effects of habitat-specific competing mortality risks on population abundance and on the sex and age-class structure. Similar but simpler stochastic simulation models have been developed for the southern Great Barrier Reef green (Chaloupka, 2002a) and loggerhead (Chaloupka, 2003a) sea-turtle populations. The southern Great Barrier Reef green turtle model presented in Chaloupka (2002a) was extended (Chaloupka, 2004) to account for a simple meta-population structure based on distance-dependent dispersal. Mazaris et al. (2009) developed an individual-based stochastic simulation model that accounted for various density-dependent biological and behavioral attributes (e.g., nest-site selection) of nesting loggerhead sea turtles in the eastern Mediterranean. The model was designed to evaluate the potential effect of nesting habitat loss due to coastal development and sea-level rise on hatchling production and population dynamics. Similar individual-based stochastic simulation models have been used by Mazaris and colleagues to evaluate various risk factors, such as ageclass-specific mortality on nesting Mediterranean loggerhead population dynamics (Mazaris et al., 2005, 2006).

Integrated Population Dynamics Models

Stochastic simulation models outlined above are the most comprehensive models developed so far to explore the population dynamics of sea turtles and to evaluate the potential effects of exposure to anthropogenic hazards on those populations. It is possible to fit the process-based models developed, for instance, by Chaloupka (2003b) to a variety of ageclass-specific abundance and demographic data. This modeling approach, comprising integration of various data and model components and simultaneous estimation of values for all parameters, presents a number of challenges, including the availability of long time series (Fonnesbeck and Conroy, 2004). Maunder (2003, 2004) presents an integrated population-modeling framework applied in recent fishery stock assessments that warrants further investigation for sea-turtle population assessments when suitable data series exist. A similar approach was used by Fonnesbeck and Conroy (2004) to model the effects of harvesting on black-duck populations.

Multispecies and Ecosystem Models

Sea turtles interact directly and indirectly with other species, and changes in environmental factors have effects on vital rates (see Figure 3.1). There has been an increasing effort to incorporate multispecies and ecosystem interactions in fishery-assessment models (Plagányi, 2007), and any mechanistic model of sea-turtle dynamics has to account for changes in prey, predators, competitors, and habitat. However, comprehensive ecosystem models include a large number of parameters and uncertain interactions; thus, they may prove to be more heuristic than predictive (Fulton et al., 2003). Qualitative approaches, such as loop analysis of community models, can evaluate stability and trophic responses in data-poor systems (Dambacher et al., 2003). Biomass-balance models, such as EcoPath with Ecosim, require more information on food-web structure and energy transfer but have been applied to a number of ecosystems that include sea turtles (Walters et al., 1997). Comprehensive tools for ecosystem-based fishery assessment, such as Atlantis (Fulton et al., 2005), may have future application to sea-turtle management in well-studied ecosystems.

Bayesian Belief or Probability Network Models

There are few robust tools available to assist risk assessment and policy development in data-poor and knowledge-vague situations. One approach to support better decision making in data-poor situations is to apply a method known as Bayesian belief networks, also known as probability networks or Bayes nets (Varis and Kuikka, 1999; Castelletti and Soncini-Sessa, 2007). It provides a structured framework to integrate information from several sources, including simulation models, published material, and stakeholder and expert opinion. Chaloupka (2007) introduced this probability-based approach at a recent workshop of the Food and Agriculture Organization of the United Nations as a robust way to evaluate the relative risk of effects of ageclass-specific anthropogenic hazards—such as fishing gear, coastal development, and climate change—on the long-term viability of Southeast Asian sea-turtle populations. The Bayesian belief network model constructed for the workshop showed (given limited data and uncertainty about turtle-fisheries interactions) that trawl fisheries, gillnet fisheries, and coastal development were hazards most likely to have major effects on the viability of the Southeast Asian sea-turtle populations.

Potential Biological Removal

Some models are designed specifically to address particular management questions, such as identification of a threshold bycatch level (see Table 6.1). Potential biological removal (PBR) was developed for marine-mammal populations to determine a maximum removal rate that a population can absorb without a large increase in the probability of decline (Barlow et al., 1995; Wade, 1998). PBR is based on the precautionary approach in a very explicit way. A simple algebraic formula is based on the concept of optimum sustainable yield (Taylor et al., 2000), which is a function of population productivity. PBR determines a maximum human-caused removal of individuals from a population on the basis of half its potential net productivity rate, adjusted by a recovery factor (F) that varies from 0.1 to 1 depending on the status of protection. The equation requires a minimum population estimate (N_{min}), the maximum rate of increase predicted (or measured) for a population (R_{max}), and predetermined risk criteria (low risk to minimal risk) for the recovery factor. PBR is generally applied to an entire population or stock but could be set for specific life stages; the PBR value represents cumulative removals due to all anthropogenic sources. PBR and various modifications to accommodate sea-turtle life history have been explored (Bolten et al., 1996; Turtle Expert Working Group, 2000) but not yet used to set bycatch limits or evaluate human-caused mortality.

Each of those modeling approaches has merit in potential application to sea-turtle demographic analysis and assessment. However, no model can be useful without data for both setting values of parameters and evaluating model behavior, particularly for applications that require precision. Increasing model complexity provides biological realism and the ability to estimate population status precisely, but data need to increase also (see Table 6.1). The most biologically realistic and complex models for sea turtles have been developed for populations with long time series of in-water abundance, breeding frequency, survival-rate estimates, and nesting abundance (e.g., Chaloupka, 2003a, b). All of the published sea-turtle assessment reports (e.g., the Turtle Expert Working Group reports and National Marine Fisheries Service [NMFS] technical memoranda summarized in Table 1.2) have highlighted the paucity of basic data for population modeling, as have reviews of sea-turtle modeling efforts in the United States (e.g., Heppell et al., 2003). The most recent sea-turtle status assessments (National Marine Fisheries Service and U.S. Fish and Wildlife Service, 2007a, b, c, d, e, f) also comment on the need for basic information on population structure and vital rates to identify changes in populations and their listing designations properly.

ASSESSMENT PROCEDURES FOR SCIENTIFIC REVIEW OF DATA AND MODELS

In addition to identification of appropriate assessment tools, it is important to have standard procedures for evaluation that ensure rigorous scientific review in all phases of assessment. A thorough review process that covers all elements of a stock assessment is invaluable when it is undertaken by knowledgeable teams of scientists that also include independent experts. It ensures that the "best available science" (National Research Council, 2004; Sullivan et al., 2006) is used to manage our nation's resources, especially when the process is transparent and open to the public. The need for "best available science" is encoded in legislation directly applicable to sea turtles under the Endangered Species Act of 1973 and in Standard 2 of the Fishery Conservation and Management Act of 1976 (reauthorized in 1996 as the Magnuson-Stevens Fishery Conservation and Management Act). To achieve the use of best scientific data and practice, assessments may include several components that each include peer review.

Fishery Assessment

The review procedures for stock assessments vary regionally in the United States depending on the fishery management council that is responsible for managing the stock, but they follow a general pattern wherein panels of experts review input data series, models, and reference points. The review workshops are the Stock Assessment Workshops and the Stock Assessment Review Committees in the northeastern United States; the Southeast Data, Assessment, and Review in the southeast and Gulf of Mexico regions; and the Stock Assessment Review and the Western Pacific Stock Assessment Review in the Pacific region. Typically, the expert panels include one or more members of the management council's Scientific and Statistics Committee (SSC) and state, federal, and academic scientists but may also include international reviewers from the Center for Independent Experts (CIE). The reviews entail workshops that last up to a week; the workshops result in a series of written reports that are available through the NMFS Web site. CIE reports are prepared separately and are also available to the public online. The CIE reports provide an independent and critical review that are outside agency procedures or oversight and can provide valuable insights. The process of fishery assessment and formulation of management recommendations involves a series of workshops.

Data Workshops

Participants in data workshops are experts who are responsible for data programs and collections. Some data-review workshops also include CIE representatives who evaluate data quality and the statistical analyses used in data summaries. During the data workshops, input data are submitted by state agencies and NMFS that include (1) fishery-dependent measures, such as catch per unit effort (CPUE), total catch, and age-length matrices to convert total catch to catch at age among others; (2) fishery-independent measures, such as survey catch abundance and CPUE, and biological metrics; and (3) other ancillary data that might affect abundance or distributional characteristics of the species. Those data are evaluated for consistency and quality. Data that are chosen for analysis are then recommended for use in the modeling process. Although some data that are typically used in fishery assessments are not available or directly applicable to sea turtles, the approach of comprehensive data review holds value as a potential component of sea-turtle management. It might have value in evaluating surveys, such as nesting-beach counts, strandings and in-water mark–recapture efforts, and length distributions.

Model Workshops

Participants in model workshops include assessment scientists and demographers, CIE reviewers, SSC members, and other knowledgeable experts. During the model workshops, the adequacy of input data for modeling, model performance, and stability are evaluated. In large part, the evaluations are based on the fits of model outputs to time series of population data, including abundance and age distribution. Results from several different models (e.g., biomass versus age-structured) are often evaluated after the recommendation given by the National Research Council (1998). The model results are reviewed as to whether there is evidence of sustainability of population abundance and excess mortality. Models are also reviewed for retrospective patterns in residua that indicate poor model fit as parameter values are updated over time.

Reference-Point Workshops

Participants in the reference-point workshops include experts from state and federal agencies, CIE, SSC, academics, and other knowledgeable experts. Participants evaluate the adequacy of point values that demark the level of overfishing or excess fishing mortality and the level of stock abundance or biomass that results in sustainable populations, which are sufficiently productive of new recruits. These workshops require information on population growth and productivity for evaluation of the appropriate reference points.

Management-Strategy Evaluation Workshops

The management-strategy evaluation (MSE) concept was developed in Europe and Australia to provide a simulation approach to evaluate management strategies by simulating the effects of different input data, reference points, and modeling frameworks on virtual "populations" (see, for example, Smith et al., 1999). MSE deals directly with uncertainty by simulating the entire process of population dynamics and management from data input to reference points and management response; this is the simulation version of adaptive management. The approach has many advantages because sensitivity analyses can reveal how data quality, assessment-model structure, reference points, and the management process itself affect the performance of a given management model. MSE concepts have been introduced into the stock-assessment process by CIE reviewers.

Marine-Mammal Assessment

Like sea turtles, marine mammals are protected species in the United States that face threats of mortality often caused by direct and indirect interactions with fisheries. Section 117 of the Marine Mammal Protection Act specifies requirements for stock assessments of marine mammals. The act requires formation and support of regional scientific review groups consisting of experts in marine-mammal ecology, population dynamics and modeling, and commercial fishing practices. The groups are responsible for reviewing stock assessments and updates and data and models used to estimate abundance and trends and for advising the agency on uncertainty and research needs. In addition, take reduction teams (TRTs), consisting of scientists and industry representatives, are formed when fishery interactions exceed the allowable take determined through PBR analysis. TRT plans are reviewed according to independent guidelines that have been established for all assessment procedures, including take evaluation, PBR calculation, and review and revision of stock-assessment reports (Wade and Angliss, 1997).

Sea-Turtle Assessment

A variety of assessments of sea turtles have been conducted by NMFS, all with considerable peer review but not as part of a standardized procedure (see Table 1.2). Sea-turtle assessments are conducted as part of a status review required by the Endangered Species Act or in response to a specific management concern. Turtle Expert Working Groups consisting of agency scientists, academics, and scientists associated with stakeholder groups have been formed at irregular intervals since 1995 to review data and conduct analyses related to conservation concerns (Turtle Expert Working Group, 1998, 2000, 2007, 2009). Status reports, required for each species every five years, are conducted by biological review teams that are composed of agency scientists in NMFS and the U.S. Fish and Wildlife Service (USFWS); these are primarily data-update summaries but recently included quantitative analysis (Conant et al., 2009). Recovery teams update the recovery plans for each species, which are split by ocean basin; recovery plans use existing models or published model results to set recovery criteria (e.g., National Marine Fisheries Service and U.S. Fish and Wildlife Service, 2008). Expert workshops to evaluate particular assessment-related issues, such as survey techniques and fishery-impact assessment, have been conducted with assistance from academic scientists and fishery management councils (e.g., Bolten et al., 1996). In addition, the agency conducts and contracts out quantitative evaluations that result in internal agency reports, including take evaluations described in biological opinions. All documents are submitted to extensive internal review and various degrees of external review. Recent quantitative analyses used by Turtle Expert Working Groups, biological review teams, and recovery teams have undergone external review by CIE.

Take Evaluation

In accordance with Section 7 of the Endangered Species Act, all federally permitted activities that have potential interactions with sea turtles are evaluated for effect. Estimates of the number and severity of interactions with sea turtles are developed by using observer data or other sources. The resulting biological opinions include the population level impacts of takes, where a take may be direct or indirect killing, injuring, or harassment of individuals or their habitat. Activities may need to be reduced or restricted if they are likely to impede recovery of a listed species or stock. For sea turtles, which are under the joint jurisdiction of NMFS and USFWS, biological opinions are most often written in response to sea-turtle interactions with commercial fisheries or coastal development activities. Under Endangered Species Act guidelines, the evaluations must include a determination of whether a proposed activity is likely to cause "jeopardy" to the affected population or species as a whole. Biological opinions and jeopardy rulings are critical documents in litigation and are challenged regularly by environmental and industry groups. Standardized, quantitative tools are desirable to determine when a take is sufficient to cause jeopardy and to warrant a curtailment of the fishing or development activity. PBR was developed for marine mammals for a similar application (Taylor et al., 2000).

Quantitative evaluation of the effects of bycatch on sea-turtle recovery has been discussed in workshops (Bolten et al. 1996) and modeled in various ways by expert working groups (Turtle Expert Working Group, 2000), agency scientists (National Marine Fisheries Service Southeast Fisheries Science Center, 2001; Snover, 2008), and contractors. In all cases, the authors blamed a lack of basic demographic information for their inability to discriminate among alternative models. The uncertainty in past and present survival, growth, and reproduction rates was too high to make a proper assessment of the likely effect of bycatch at the population level. In one case, 64 alternative population-projection scenarios for loggerheads were presented, ranging in prediction from dramatic decline to rapid recovery (National Marine Fisheries Service Southeast Fisheries Science Center, 2001). A more complex evaluation of expected changes in population growth that might result from reductions in anthropogenic mortality used age-structured models with Monte Carlo sampling of vital-rate distributions to try to cope with uncertainty; the result was a nearly incomprehensible amalgamation of possible population responses (National Marine Fisheries Service Southeast Fisheries Science Center, 2009). Without demography, there is no way to predict the effects of fishery bycatch for such a long-lived animal (Heppell et al., 2003).

Threats Evaluation

Recent recovery plans have included a semiquantitative evaluation of threats to sea-turtle populations based on rough estimates of the number of turtles affected. To compare the potential population-level effects of threats that affect different life stages of sea turtles, the recovery teams have developed an "adult equivalent" calculation that "discounts" the estimated number of juvenile mortalities according to their reproductive value relative to the reproductive value of adults (National Marine Fisheries Service and U.S. Fish and Wildlife Service, 2008; Wallace et al., 2008; Bolten et al., in press). Reproductive value is determined by a deterministic-lifecycle matrix, which requires estimates of survival, growth, and fertility. Uncertainty in remigration interval or other reproductive parameters can have a substantial effect on the adult reproductive value used for scaling, and reproductive values depend on the underlying asymptotic growth rate predicted by the matrix (Caswell, 2001). Thus, methods based on reproductive value and adult equivalents rather than quantitative assessment of threats or setting take limits are best for relative comparisons within species that may be used to set priorities for research or conservation effort.

Abundance Estimation

Estimating population size of sea turtles is highly problematic because they inhabit vast areas and have many ageclasses that occur in different habitats. Extrapolation of nest abundance and trends to adult sea turtles, which probably make up less than 5% of the non-hatchling population

(Crowder et al., 1994), requires data on sex ratio, recruitment rates (proportion of nesters that are breeding for the first time), and annual survival; uncertainty in these parameters has been incorporated through resampling of known or presumed distributions to provide a range of possible population sizes (Turtle Expert Working Group, 2007). Extrapolation of nesting data to estimate population size is even more problematic because of uncertainty in survival and cohort variability. A lack of sufficient information on survival rates resulted in a range of a factor of five to ten in estimates of population sizes among best-fit models for Kemp's ridley sea turtles even though cohort strength (annual hatchling production) was well known on the basis of extensive monitoring of nests for the entire species (Turtle Expert Working Group, 2000; Heppell et al., 2005).

Population Trends and Probability of Extinction or Recovery

Older sea-turtle assessments relied heavily on simple regression analysis of nesting-beach data to evaluate population trends, but recent assessments published by NMFS have included Bayesian state-space modeling and diffusion approximation methods to estimate trends and uncertainty in population trajectories (Turtle Expert Working Group, 2007, 2009; Conant et al., 2009). The most recent status assessment of Atlantic loggerhead turtles also includes a "matrix threat analysis" that is essentially a deterministic matrix sensitivity analysis to ascertain potential changes in population growth that result from additional mortality (Conant et al., 2009). The analysis is far more comprehensive than past sensitivity analyses (e.g., Crowder et al., 1994; Heppell et al., 2003) in that it accounts for uncertainty in estimates of parameter values. The potential cumulative effects of anthropogenic stressors affecting all life stages of each population unit are then modeled as additive mortality, and ranges of potential asymptotic growth rates are compared. The exercise is informative insomuch as it shows that even under the most optimistic scenarios, there is a high probability that current mortality is too high to be sustained by most loggerhead populations. However, it is a largely heuristic exercise with little or no real power for prediction because of the high level of uncertainty and of assumptions required for deterministic age-structured models. There is no attempt to fit models to data, in part because the time lags involved in sea-turtle life history make it very difficult to establish a likely past or current age structure of the population.

CONCLUSIONS

Population assessment for management requires an integration of abundance data and demography to account for species' life history and to determine the likely causes of observed trends. There are a number of modeling approaches of varied complexity and precision that can address management questions, but they all need accurate data at the population level. Vital-rate estimation is essential for these slow-growing species, as trends in nesting-beach abundance provide information about only a tiny fraction of a sea-turtle population. Some data that can be used to determine changes in vital rates already exist, including time series of juvenile abundance (or indexes of abundance) and size distributions.

Assessments of managed fish populations include gathering and reviewing biological information and catch data, a variety of modeling workshops to determine the most appropriate tools for assessment and reference points for status determination, and extensive external peer review. Marine-mammal assessments also follow a prescribed path for evaluation. Sea-turtle assessments have included many of the elements required for those species but are not done in a set procedural framework that ensures consistency, transparency, and thorough evaluation.

There has been no thorough attempt to assess sea-turtle status with population models that are fitted by using available data on bycatch, size distributions, and productivity. That is because of the following three primary factors that can be addressed by the agency:

- Critical vital rates have not been monitored so there is high uncertainty in estimates of parameter values and in interpretation of trends.
- Data are scattered and require a thorough evaluation to determine their quality and their applicability to population assessment.
- Sea-turtle assessment efforts have not been isolated from broader evaluations of status and threats and have rarely included scientists in other quantitative modeling fields, such as fishery scientists.

RECOMMENDATIONS

- NMFS and USFWS should develop a general framework for a sea-turtle assessment procedure, including data evaluation, model review, and MSE.
- NMFS and USFWS should conduct data-evaluation workshops, starting with Atlantic loggerheads, focused specifically on the evaluation of time series information that can contribute to setting values of parameters for demographic models. Data for evaluation include, but are not limited to, nesting abundance, in-water abundance, hatchling-cohort production, length distributions, and reproductive frequency. All sources of data should be evaluated for quality, consistency and spatial or temporal heterogeneity, and gaps.
- Researchers should work with modelers in different fields to

develop a toolbox for sea-turtle assessment that can provide standardized methods for evaluation and review of data-poor and data-rich species. They would include methods that use available data on trends and size distributions of turtles to reduce the possible ranges of unknown values of parameters and estimates of abundance through model fitting.

- The agencies should sponsor a cost-benefit analysis workshop to set priorities among research needs according to which parameters will provide the most useful information for diagnosis of population change.

7

Cross-Cutting Issues: Data, Education, Permits, and Coordination

DATA MANAGEMENT

This report describes extensive data requirements for accurate assessments of sea-turtle populations. Many of the required data have yet to be collected. Other data resources currently exist but have not been used to address data gaps because the data are not accessible or have inadequate access and are at risk of being lost as data owners change fields, retire, or die. Most of those datasets cannot be replaced because they were collected in past years or even decades under different environmental conditions and turtle densities.

The situation is not unique to sea turtles. The need for open access to data has been recognized for decades, and many committees and workshops have been convened to discuss and develop methods to address the need. A recent National Research Council (2009) report, *Ensuring the Integrity, Accessibility, and Stewardship of Research Data in the Digital Age,* cites 36 reports of the National Academy of Sciences, the National Academy of Engineering, the Institute of Medicine, and the National Research Council published since 1985 that have addressed the issue. A recent editorial (Whitlock et al., 2010) by editors of four prestigious scientific journals emphasized the need for archiving raw data—not data summaries—to prevent the loss of critical data to science and announced a new policy. Several journals of ecology and evolution will now require authors to submit to an appropriate public archive all the raw data on which their articles are based.

Three critical issues emerge (National Research Council, 2009): data integrity; data access, sharing, and ownership; and data stewardship and management plans. The National Research Council report concluded that explicitly outlining the roles and responsibilities of the various entities—data providers, host institutions, and data users—is essential.

In this report, the committee has not repeated information that was so thoroughly reviewed elsewhere. Rather, it has described the current situation for sea-turtle data and has recommended what should be done to make data accessible for research and management and to reduce the risk of data loss.

Current Status

The fractured status and lack of coordination of sea-turtle databases are major impediments to the management and conservation of sea turtles. Throughout the United States, hundreds of projects (of varied duration) have been established to monitor sea-turtle populations and conduct research on sea-turtle biology. The projects have been conducted by people in federal and state agencies, universities, and nongovernmental organizations (NGOs) and by private individuals.

Data resulting from those projects have varied integrity, accessibility, and stewardship. The integrity (structural completeness, including metadata [data that provides information about other data]) and quality of the data vary greatly, depending on many factors. Quality control of data collection is a major factor. Factors affecting data-collection quality include the extent and consistency of training given to data collectors; the experience and number of data collectors; and the quality of equipment used, such as tags and instruments to measure turtles. Transcription of data from field or laboratory notes to digital databases is a common source of errors. Quality control of data transcription is essential to maintain the integrity of the database. Accurate metadata can help to offset some data-quality problems. For example, accurate reporting of annual survey efforts can offset problems of uneven effort among years. An important difficulty in sea-turtle count data is understanding whether each zero count is the result of the absence of turtles or the result of zero effort.

Many databases resulting from sea-turtle studies have little or no access for people other than the data owners. Data accessibility is determined by the willingness of a data owner to share the data, the ease of data use, and the presence of essential metadata so that data can be interpreted. Some data, particularly those from federal and state agencies, are available as digital databases but only in summary form.

Current stewardship of the data resulting from the projects ranges from well-curated, computerized databases with safeguarded backups to boxes of loose data sheets stored in a single vulnerable location. Data from some studies have been lost and cannot be reconstructed. The risk of loss of the databases depends on a number of factors, including the form of the data, arrangements for perpetual management, and the number of people and type of organization involved.

Examples of Sea-Turtle Databases Established to Share Data

Some databases provide information to locate data sources or to avoid duplication and confusion in sea-turtle studies. Because that type of database does not threaten "ownership" status, participation tends to be good, depending largely on the benefit to participants, which varies from certain and immediate (e.g., Marine Turtle DNA Sequences) to less certain and future (e.g., the Sea Turtle Tag Inventory). Some examples of such databases are noted below:

- *In-Water Sea Turtle Monitoring and Research in Florida: Review and Recommendations* (Eaton et al., 2008) lists all 42 known in-water sea-turtle research projects in Florida (active and inactive) with maps, brief summaries of results, and lists of publications. Given funding opportunities, the Florida Fish and Wildlife Conservation Commission has tentative plans to update this database and initiate coordination among projects.
- Marine Turtle DNA Sequence Web sites assign haplotype (i.e., nucleotide sequence) designations to all mitochondrial DNA sequences for green turtles (*Chelonia mydas*) and loggerhead turtles (*Caretta caretta*) in the Atlantic (University of Florida, 2001) and Pacific green turtles (Southwest Fisheries Science Center, 2008) as they are discovered to facilitate coordination and to avoid the confusion of duplication of sequence designations in publications. The databases are updated as new sequences are submitted.
- The Sea Turtle Tag Inventory (University of Florida, 1999a) lists all flipper-tag sequences used by programs around the world to avoid duplication of tag numbers when tags are purchased by different research programs and to assist in reporting data on the recapture of turtles when only the tag number has been recorded. The database is updated as new tag series are submitted.

Several databases secure data from sea-turtle projects (protect them from loss) and provide partial access or access to data summaries. However, long-term continuation of the host institution is not always ensured. A few examples are the following:

- The Cooperative Marine Turtle Tagging Program (University of Florida, 1999b) is a centralized program funded by the Southeast Fisheries

Science Center of the National Marine Fisheries Service (NMFS) and managed by the Archie Carr Center for Sea Turtle Research at the University of Florida to distribute sea-turtle tags, manage tagging data, correspond with people who capture tagged turtles, and facilitate exchange of tag information in the Atlantic. All data owners allow NMFS to use their data for management purposes and stipulate any additional extent to which their data are accessible. All data owners allow the Archie Carr Center for Sea Turtle Research to release original tagging data to people who report the capture of tagged turtles.

- Satellite tracking (Seaturtle.org, 2009) displays maps of sea-turtle tracks generated by satellite telemetry with contact information on data owners. Raw data are not available, and use of data is not allowed without permission from data owners.
- Sea-turtle nest-count data for Florida (Florida Fish and Wildlife Conservation Commission, 2009) are displayed in a summary table of the statewide nesting totals for each year beginning in 1979 for each species and for the most current year (updated each February) only as a summary of nests and nonnesting emergences by county for each species.
- Data on sea turtles that stranded along the coast from Maine to Texas in 1998–2005 (Southeast Fisheries Science Center, 2010) are available from as monthly totals for each species for each county. Beginning in 2006, data are available as weekly totals for each species for each NMFS zone, divided into inshore and offshore categories.
- Nesting and stranding data on a few areas are available from Seaturtle.org (2010a, b).

Few databases secure the data and provide complete access to raw data. The following describes the two major types of data collected (tagging data and geographic distribution and abundance data):

- The Legacy Database Initiative of the Archie Carr Center for Sea Turtle Research (University of Florida, 2010) will consist of many datasets. The first (completed) contains the tagging data on nesting sea turtles at Melbourne Beach, Florida, collected by Billy J. Turner and colleagues in 1972–1981. The second will contain the tagging data on nesting sea turtles on Jupiter Island, Florida, collected by Frank Lund and colleagues in 1969–1981. All data, with accompanying metadata, will be available on the Web.
- The Ocean Biogeographic Information System Spatial Ecological Analysis of Megavertebrate Populations (OBIS-SEAMAP; Duke University, 2009) is a spatially referenced online global database on megavertebrate (including sea-turtle) distribution. The database can be searched and viewed through online mapping applications. Raw data are available

to download under the agreement that data contained in OBIS-SEAMAP will not be used in any publication, product, or commercial application without prior written consent of the original data provider. OBIS-SEAMAP is in OBIS, which was established by the Census of Marine Life.

Centralized Data Facility

Perhaps the most efficient and secure approach for making sea-turtle data accessible would be to have all databases available through a single, permanent facility that would ensure long-term management of the data. For tagging data, one possible program is the Cooperative Marine Turtle Tagging Program, described above. For geographic distribution and abundance data, a possibility is the Global Biodiversity Information Facility (GBIF), an international organization whose goal is to make biodiversity data accessible everywhere in the world. OBIS is an associate member of GBIF, which is the only intergovernmental organization mandated to make data on global biodiversity freely accessible. It is now the largest, most comprehensive portal to biodiversity information with more than 177 million biodiversity data records (Gilman et al., 2009).

Top priority needs to be given to coordinating data from within the United States and its territories. That effort would require extensive coordination among federal and state agencies, NGOs, and individual citizens. Because sea-turtle populations are shared by many nations, concerted efforts could also be made to coordinate with governments and NGOs in other countries. International networks, such as the Wider Caribbean Sea Turtle Conservation Network, could be valuable partners.

Data Protocols

Consistent data collection maximizes the ability to combine and compare data among studies. Attempts have been made to standardize protocols for collection of data on sea turtles by a wide range of techniques (e.g., Bjorndal and Balazs, 1983; Higgins et al., 1997; Eckert et al., 1999; National Marine Fisheries Service Southeast Fisheries Science Center, 2008). Because many sea-turtle research programs have been underway for extended periods, it is understandable that researchers would be reluctant to change or add methods to their own data protocols.

Archives

In addition to the data archives discussed above, there is a great need for archives to store tissue samples from sea turtles. An archive for genetics samples already exits at the NMFS Southwest Fisheries Science Center, and the National Oceanic and Atmospheric Administration Center for Coastal Environmental Health and Biomolecular Research in Charleston, South Carolina, maintains a small archive of sea-turtle tissues. Additional archives are needed for various tissue types to support analyses of somatic growth through skeletochronology, of resource use through stable isotope analyses, and of contaminant loads through analyses of pollutants. The archives need to be curated carefully and provide long-term storage and access to researchers. Incentives in the form of analytical assistance, collaboration help, and facilitated access will be needed to maximize contributions to archives.

EDUCATION AND CAPACITY BUILDING

Chapter 1 emphasizes the need for U.S. management agencies to apply a more complete and quantitative understanding of sea-turtle population dynamics to management policy. Limits on quantitative information pertinent to sea-turtle management stem from both inadequate quantitative expertise and insufficient guidance of study designs and data analyses of policy information needs. Short-term remedies for the problem might include recruiting statistics and modeling specialists into management agencies from fields outside conservation biology. However, effective analyses applied to pertinent management needs would require difficult science and policy translation between fields. That translation has been especially difficult in the interpretation of scientific uncertainty (Bradshaw and Borchers, 2000) and in the use of caution in management decisions (Cooney, 2004). The present committee proposes that long-term remedies for pertinent quantitative information deficits include interdisciplinary training of fishery and conservation professionals. (See the discussion of assessment procedures in Chapter 6.)

There is interest both nationally (Jacobson and Robinson, 1990; National Research Council, 2000) and internationally (Buitrago et al., 2008) in the interdisciplinary challenges of educating quantitative fishery and conservation professionals. There is general agreement that the education of effective professionals needs to be broadly based and interdisciplinary (Massey, 1989; National Research Council, 1998; Clark, 2001). Training needs to include both quantitative and biological subjects, such as population and ecosystem ecology, statistics, and modeling; but it also needs to include economics, policy, and decision-analysis courses, for example, to provide insight into how conservation of natural resources can be achieved. Students are generally eager to take fundamental biology courses, such as physiology and anatomy, but commonly avoid the fundamental courses in mathematics and statistics that are needed to establish sufficient quantitative skills. Although there is general agreement, realization of interdisciplinary education faces structural barriers in colleges and universities (Jacobson, 1990; Jacobson and Robinson, 1990).

Capacity building includes the development of partnerships between government and NGOs. With respect to sea-turtle conservation and management, they would include NMFS, the U.S. Fish and Wildlife Service (USFWS), the U.S. Geological Survey, state resource agencies, universities, aquariums, NGOs, biology consultants, and international collaborators. In addition to the development of partnerships, capacity building includes public outreach and improvement of scientific infrastructure.

An example of the difficulty in improving human resources for conservation work can be found in a report by the U.S. Department of Commerce and the U.S. Department of Education (2008). The fishery-management and marine-conservation agencies face the same challenges that were identified in a National Research Council (2000) report on recruiting quantitative scientists to the agency despite aggressive actions to provide educational opportunities. More than 15 U.S. universities are engaged in cooperative programs with NMFS, along with Sea Grant-administered Graduate Fellowships in Population Dynamics and Marine Resource Economics. Other programs, such as the NMFS–Recruiting Training Research Program at Virginia Polytechnic Institute and State University, undertake special population-dynamics workshops annually for undergraduates. Despite those advances, conservation education faces the challenges of providing interdisciplinary education within a traditional academic framework (Le Tissier et al., 2004; Kroll, 2007).

Important quantitative elements of interdisciplinary training for conservation students include a working knowledge of basic models and statistical evaluation of data. Students need to be aware of quantitative analysis' value as a provider of recommendations and as an evaluator of potential sources of bias and uncertainty and key evidence for motivating conservation action. All students who major in natural or social sciences require an understanding of models and population effects of management actions if they are to evaluate critically and correctly the tools that are used in decision making.

To promote a broader appreciation of the uses, and potential misuses, of quantitative analysis, universities need to make population-dynamics training more widely accessible to undergraduates (Hard, 1995; Matter and Steidl, 2000; Burger and Leopold, 2001). Existing biology and natural-resource programs sometimes leave out population-dynamics education beyond basic theoretical models presented in ecology or leave a gap between basic, introductory courses and highly technical quantitative courses for graduate students.

ALLOCATION OF MANAGEMENT AND RESEARCH FUNDS

Federal agencies need to ensure that funds available to support research—both internal and external funds—are invested wisely. At a minimum, all research proposals generated in federal agencies have to be reviewed by panels that include federal and nonfederal scientists.

An example of one system that is working well is the Western Pacific Regional Fisheries Management Council (WPRFMC) Sea Turtle Conservation Program. The committee summarizes its approach here, not as a description of what should be done but as a starting point for agency-appropriate plans. The program was established in 2002 to ensure the sustainability of Hawaii-based longline fisheries, contribute to the international transfer of sustainable fishery technology and knowledge, and aid in the recovery of Pacific sea-turtle populations. Since then, WPRFMC has played an instrumental role in fostering collaboration, transferring bycatch-mitigation technology, and advancing the sustainability of fisheries by convening a number of international meetings. It has also played a key role in encouraging sea-turtle research, monitoring, and conservation projects in the Pacific where funding may not have been otherwise available, and its program annually receives a portion of the congressional funding dedicated to Pacific sea-turtle research and conservation. With the advice of its Sea Turtle Advisory Committee (STAC), WPRFMC has been supporting conservation measures since 2003 to offset adverse effects on sea-turtle populations from the Hawaii-based longline fishery. STAC was formed by WPRFMC at the 114th council meeting (August 2002) to direct and advise on its turtle-conservation activities. STAC generally meets once a year and comprises eight well-known sea-turtle biologists and scientists. In FY 2010, WPRFMC initiated an annual unified request for proposal (RFP) process for WPRFMC-funded sea-turtle conservation projects. The RFP process solicits projects focusing on one or more of WPRFMC's high-priority species and activities, as defined by its five-year plan for 2010–2014 and recommended by STAC. Proposals are reviewed by a panel consisting of WPRFMC staff, STAC members, and additional external reviewers if necessary. All previously funded projects requesting continued support from WPRFMC are subject to annual review through the same RFP process.

PERMITTING

Before initiating a research project on sea turtles in the United States that has potential for take, investigators must obtain one or more research permits. The NMFS Office of Protected Resources is responsible for permitting studies conducted in the water, and USFWS is responsible for research conducted or initiated on land. For example, USFWS would issue a permit to attach a satellite transmitter to a sea turtle that has come ashore to nest and will return to the ocean. USFWS has established cooperative agreements with states and territories (Florida, Georgia, South Carolina, North Carolina, Puerto Rico, and the U.S. Virgin Islands) that have dedicated staff with sea-turtle expertise so that state agencies may grant permits under the auspices of Section 6 of the Endangered Species Act for research and educational programs on threatened sea-turtle species. NMFS does not have a similar relationship with states so a sea-turtle research project that is to take place in state waters usually requires permits from both NMFS and the state in which the work is to be conducted, although in some cases the federal permit is all that is required.

The committee has found broad consensus among researchers studying sea turtles that the permitting process is a greater obstacle to research than is necessary for the protection of the turtles or for meeting the requirements under the Endangered Species Act. There are three major concerns. First, the process is too slow; permits often take six months and sometimes much longer to be issued or denied. Second, the review is redundant and sometimes inconsistent with other required reviews, such as those rigorously conducted by internal and external scientific peer review of proposed research by funding agencies and by Institutional Animal Care and Use Committees; there is also redundancy between federal and state or territorial agencies that have federal permitting authority. Third, the reasons for rejecting a permit request are not always provided, and mechanisms for appeals are not specified.

New research projects with innovative techniques will need to be initiated to meet data needs outlined in this report. However, in numerous examples presented during committee meetings, the U.S. permitting process delayed or denied research projects, particularly when innovative techniques were involved. The permitting process need not unnecessarily delay or hamper these critical studies. Considering the balance between overregulation and underregulation, it is clear that the sea-turtle research-permitting process is not underregulated. Evidence of that is the absence of third-party lawsuits challenging granted research permits, whereas numerous lawsuits have resulted from the issuance of nonresearch incidental-take permits. Permitting agencies need to improve efficiency and change research-permitting processes so that the Endangered Species Act mandates are met in a timely and transparent process for permit applicants.

RECOMMENDATIONS

Data Management

- To avoid the overlooking of data sources, NMFS should create a metadatabase[1] identifying as many of the sea-turtle datasets in the United States and its territories as possible, similar to the document created for in-water projects in Florida. The online database should be updated regularly. As was done with the Florida in-water project, the permits granted for monitoring and research through federal and state agencies can be used to identify many of the projects. The database would provide information on available data, status of each dataset (e.g., computerized, hard-copy only, lost), and contact information but would not include the data.
- NMFS and USFWS should partner with other government agencies, universities, and NGOs to develop a mechanism to obtain, computerize, maintain, and make accessible as many sea-turtle databases as possible. There is some urgency in this task while data collectors are still available to provide essential metadata. Such issues as data ownership, authorship requirements, and ensuring appropriate use of data will need to be addressed through data safeguards, extensive outreach, and participant incentives. Priorities for selecting which databases to conserve should be based on the integrity of the data, the amount and type of data, and risk of loss.
- NMFS and USFWS should partner with other government agencies, universities, and NGOs to improve coordination among data holders. Incentives should be developed to encourage data sharing; these may include providing participating researchers with data-analysis services and data products, regional data summaries, data backup assurance, assistance with publication of results, and facilitation of collaborative relationships.
- The Sea Turtle Stranding and Salvage Network should make information on all stranded turtles available for evaluation at least by review teams and assessment modelers.
- NMFS and USFWS should convene a working group of experts in government agencies, academia, and NGOs to consider establishing centralized databases of all sea-turtle data collected in the United States and its territories.
- NMFS and USFWS should convene a task force of experts in government agencies, academia, and NGOs to establish standard research

[1] A metadatabase manages data that provide information about other data or are derived from other data.

and data-collection protocols, building on earlier work (Eckert et al., 1999; National Marine Fisheries Service Southeast Fisheries Science Center, 2008), with emphasis on techniques that have recently emerged. The task force should also develop incentives for researchers to adopt the protocols and outline a plan for continuing training in methods and analytical techniques.

- NMFS and USFWS should establish and maintain long-term tissue banks, similar to the genetics tissue bank now at the Southwest Fisheries Science Center, for other types of tissues. The agencies should develop effective incentives to encourage participation in tissue banking, such as collecting humeri from turtle carcasses and tissue samples from turtles captured incidentally in fisheries.

Education and Capacity Building

Most of the recommendations presented in an earlier report of the National Research Council (2000) are still relevant today, and many of them remain unfulfilled. In addition, the present committee recommends the following:

- Increase opportunities for undergraduates to have hands-on experiences with sea-turtle conservation and population dynamics. That could be done by increasing funding of existing cooperative programs or by developing summer programs similar to the National Science Foundation's Research Experience for Undergraduates. Because quantitative skills are essential for species management generally, summer courses could be directed toward a broader audience of undergraduates and beginning graduate students who are pursuing careers in conservation of marine mammals, sea birds, and other marine species.
- Increase opportunities for graduate and postgraduate students to pursue quantitatively oriented careers in conservation biology. That could be accomplished by funding additional scholarships in the NMFS–Sea Grant Joint Graduate Fellowships in population dynamics and in marine-resource economics.
- Provide support for hands-on workshops that include (1) introductory materials (in English and Spanish) that provide a basic overview of why quantitative evaluation and statistical rigor in data collection are important for sea-turtle conservation and (2) data analysis and modeling guidance on how to analyze data and interpret model results. The workshops would be valuable tools in connecting data holders across regions.
- Provide outreach and training in how scientific information shapes conservation policy.

- Expand and facilitate involvement in student internships in the NMFS Office of Protected Resources.
- Formalize an outreach program aimed at informing professionals in conservation biology about how the information that they gather is used in management decisions. Clearly broadcast updated information and data needs for assessments of risks and population viability.

Allocation of Research Funds

- To ensure that research funds are invested wisely, NMFS and USFWS should have all research plans generated in federal agencies reviewed by panels that include federal and nonfederal scientists.

Permits

- NMFS and USFWS should convene a working group to evaluate the permitting process for research projects and develop methods to expedite the process while meeting legislative requirements and intent. Participants should include representatives of the permitting agencies and research scientists. The review should weigh unintended consequences of permitting delays and lost research opportunities, should review the potential risks and benefits to the listed species of changing permitting requirements and procedures, and should assess the extent to which scrutiny of research permits has resulted in substantial take reductions.

8

Conclusions and Recommendations

Overarching Conclusion: Although abundance estimates are critical for assessing sea-turtle populations, demographic or vital-rate parameters are critical for understanding and predicting trends in sea-turtle populations. The committee concludes that (1) in the United States, critical vital rates have not been adequately determined; (2) the most important procedural enhancements would be improved coordination in data collection and availability, a more efficient and transparent permitting process, and increased archiving of tissue samples; and (3) sea-turtle assessments have not been isolated from broader evaluations of status and threats and have rarely included scientists from other quantitative-modeling fields.

Overarching Recommendation: The National Marine Fisheries Service (NMFS) and the U.S. Fish and Wildlife Service (USFWS) should develop a coherent national strategy for sea-turtle assessments to improve the data-collection methods, data quality, and data availability and to develop a rigorous plan for external review of data and models used to assess population status and trends. The strategy would benefit from the focused attention of expert groups that include government officials, academics, and nongovernmental organization personnel. As recommended in all expert working group documents (see Table 1.2), research should emphasize vital-rate estimation (averages, annual variance, and ecological or environmental mechanisms that drive vital rates) and improvement in abundance estimates. The most serious demographic data gaps to be addressed include in-water abundance, hatchling-cohort production, survival of immature turtles and nesting females, age at sexual maturity, breeding rates, and clutch frequency.[1] More precise estimates of anthropogenic mortality are needed to evaluate impacts. All sources of data should be evaluated for quality, consistency, spatial and temporal heterogeneity and trends, and data gaps.

Detailed suggestions for improving the collection, analysis, and synthesis of data are provided at the end of each chapter of this report, and Chapter 6 describes appropriate models and procedures for assessments. Because assessments will involve different circumstances and management needs, the committee cannot recommend one standardized set of priorities for all assessments beyond its strong recommendation for a greater focus on demographic parameters. Some specific conclusions and recommendations that elaborate on the overarching conclusion and recommendation and represent the highest-priority needs are presented below.

Conclusion: Sea-turtle population assessments in the United States are based too heavily on estimates of abundance of adult females on nesting beaches. Although estimates of abundance of adult females are critical, without knowledge of accompanying changes in demographic rates at all life stages, the proximate and ultimate causes of population trends cannot be determined. Selection and evaluation of the best management options depend on an understanding of the basis of changes in population abundance.

Recommendation: NMFS and USFWS should ensure that estimates of abundance at life stages in addition to adult females are generated and that demographic rates are integrated with estimates of abundance in population assessments.

Conclusion: Inadequate information is available for population assessments because the data have not been collected or, if they have been collected, have not been analyzed or made accessible in a manner that allows them to be useful.

Recommendations:

- NMFS and USFWS should develop plans for the collection and analysis of data to address data gaps. The development should include outside experts who collect, analyze, and use the data.

[1] Clutch frequency refers to the number of clutches deposited by an individual turtle in a nesting season.

• NMFS and USFWS should present a comprehensive assessment plan and a data plan to sea-turtle biologists to facilitate effective data collection for the integrated approach and to obtain input from them on improvement of the plans.

• NMFS and USFWS, with other government agencies and funding sources, should support the collection and analysis of those data.

• To avoid the overlooking of data sources, NMFS should create an on-line metadatabase[2] that identifies as many of the sea-turtle datasets in the United States and its territories as possible and is similar to the document created for in-water projects in Florida (see Chapter 7). The database should be updated regularly.

• NMFS and USFWS should support a program to safeguard and make accessible as many sea-turtle databases as possible, past and present. There is some urgency to this task while data collectors are still available to provide essential metadata.

• NMFS and USFWS should partner with other government agencies, universities, and nongovernmental organizations to improve coordination among data holders. Incentives should be developed to encourage data sharing.

• NMFS and USFWS should arrange for a review of data now being collected under their auspices or with their support and evaluate the costs and benefits. For example, the sea-turtle stranding and salvage networks should be evaluated, perhaps with the assistance of the U.S. Geological Survey's National Wildlife Health Center.

Conclusion: Reviews of federal population assessments and research plans are not sufficiently rigorous and transparent.

Recommendations:

• NMFS and USFWS should develop a general framework for sea-turtle assessment procedures, including data evaluation, model review, and management-strategy evaluation.

• NMFS and USFWS should ensure that all research plans generated in federal agencies are reviewed by panels that include federal and nonfederal scientists. Using reviewers with quantitative skills, such as skills in population assessment and statistical analysis, is particularly important.

Conclusion: There are unnecessary obstacles to collection and analysis of critical data, including inadequate quantitative training of scientists and an inadequate process for issuing research permits.

---

[2] A metadatabase manages data that provide information about other data or are derived from other data.

Recommendations:

- NMFS and USFWS should partner with other government agencies and universities to improve the quantitative skills of people who are involved in designing, reviewing, and implementing the projects and assessments that are generated under a comprehensive assessment plan. These efforts will be short term (e.g., recruiting quantitatively skilled experts, improving the quantitative skills of current personnel) and long term (e.g., improving quantitative training of students).
- NMFS and USFWS should convene a working group to evaluate the permitting process for research projects and develop methods to expedite the process while meeting legislative requirements and intent. Participants should include representatives of the permitting agencies and research scientists. The review should weigh unintended consequences of permitting delays and lost research opportunities, should review the potential risks and benefits to the listed species of changing permitting requirements and procedures, and should assess the extent to which scrutiny of research permits has resulted in substantial take reductions.

References

Addison, D.S. 1997. Sea turtle nesting on Cay Sal, Bahamas, recorded June 2-4, 1996. *Bahamas Journal of Science* 5(1):34-35.

Aggarwal, R.K., T.P. Velavan, D. Udaykumar, P.S. Hendre, K. Shanker, B.C. Choudhury, and L. Singh. 2004. Development and characterization of novel microsatellite markers from the olive ridley sea turtle (*Lepidochelys olivacea*). *Molecular Ecology Notes* 4(1):77-79.

Aggarwal, R.K., A. Lalrumruata, T.P. Velavan, A.P. Sowjanya, and L. Singh. 2008. Development and characterization of ten novel microsatellite markers from olive ridley sea turtle (*Lepidochelys olivacea*). *Conservation Genetics* 9(4):981-984.

Akcakaya, H.R., M.A. Burgman, and L.R. Ginzburg. 1999. *Applied Population Ecology*. Sinauer Association Inc., Sunderland, Massachusetts.

Allard, M.W., M.M. Miyamoto, K.A. Bjorndal, A.B. Bolten, and B.W. Bowen. 1994. Support for natal homing in green turtles from mitochondrial DNA sequences. *Copeia* 1994(1):34-41.

Allen, M.S. 2007. Three millennia of human and sea turtle interactions in remote Oceania. *Coral Reefs* 26(4):959-970.

Allison, G.W., S.D. Gaines, J. Lubchenco, and H.P. Possingham. 2003. Ensuring persistence of marine reserves: Catastrophes require adopting an insurance factor. *Ecological Applications* 13(1):S8-S24.

Arthur, K.E., J.M. O'Neil, C.J. Limpus, K. Abernathy, and G. Marshall. 2007. Using animal-borne imaging to assess green turtle (*Chelonia mydas*) foraging ecology in Moreton Bay, Australia. *Marine Technology Society Journal* 41(4):9-13.

Avens, L., J.C. Taylor, L.R. Goshe, T.T. Jones, and M. Hastings. 2009. Use of skeletochronological analysis to estimate the age of leatherback sea turtles *Dermochelys coriacea* in the western North Atlantic. *Endangered Species Research* 8:165-177.

Avise, J.C., B.W. Bowen, T. Lamb, A.B. Meylan, and E. Bermingham. 1992. Mitochondrial DNA evolution at a turtle's pace: Evidence for low genetic variability and reduced microevolutionary rate in the Testudines. *Molecular Biology and Evolution* 9:457-473.

Balazs, G.H. and M. Chaloupka. 2004a. Thirty-year recovery trend in the once depleted Hawaiian green sea turtle stock. *Biological Conservation* 117:491-498.

Balazs, G.H. and M. Chaloupka. 2004b. Spatial and temporal variability in somatic growth of green sea turtles (*Chelonia mydas*) resident in the Hawaiian Archipelago. *Marine Biology* 145(5):1043-1059.

Balazs, G.H. and M. Chaloupka. 2006. Recovery trend over 32 years at the Hawaiian green turtle rookery at French Frigate Shoals. *Atoll Research Bulletin* 543:147-158.

Barlow, J., S.L. Swartz, T.C. Eagle, and P.R. Wade. 1995. *U.S. Marine Mammal Stock Assessments: Guidelines for Preparation, Background, and a Summary of the 1995 Assessments.* NOAA Technical Memorandum NMFS-OPR-6, National Oceanic and Atmospheric Administration, Silver Spring, Maryland.

Beaumont, M.A. 1999. Detecting population expansion and decline using microsatellites. *Genetics* 153:2013-2029.

Beavers, S.C. and E.R. Cassano. 1996. Movements and dive behavior of a male sea turtle (*Lepidochelys olivacea*) in the eastern tropical Pacific. *Journal of Herpetology* 30(1):97-104.

Beavers, S.C. and F.L. Ramsey. 1998. Detectability analysis in transect surveys. *Journal of Wildlife Management* 62(3):948-957.

Beerli, P. and J. Felsenstein. 2001. Maximum likelihood estimation of a migration matrix and effective population sizes in *n* subpopulations by using a coalescent approach. *Proceedings of the National Academy of Sciences of the United States of America* 98(8):4563-4568.

Beggs, J.A., J.A. Horrocks, and B.H. Krueger. 2007. Increase in hawksbill sea turtle *Eretmochelys imbricata* nesting in Barbados, West Indies. *Endangered Species Research* 3:159-168.

Bell, B.A., J.R. Spotila, F.V. Paladino, and R.D. Reina. 2003. Low reproductive success of leatherback turtles, *Dermochelys coriacea*, is due to high embryonic mortality. *Biological Conservation* 115(1):131-138.

Bell, C.D.L., J.M. Blumenthal, A.C. Broderick, and B.J. Godley. 2010. Investigating potential for depensation in marine turtles: How low can you go? *Conservation Biology* 24(1):226-235.

Benson, S.R., K.M. Kisokau, L. Ambio, V. Rei, P.H. Dutton, and D. Parker. 2007. Beach use, internesting movement, and migration of leatherback turtles, *Dermochelys coriacea*, nesting on the north coast of Papua New Guinea. *Chelonian Conservation and Biology* 6(1):7-14.

Berry, O., M.D. Tocher, and S.D. Sarre. 2004. Can assignment tests measure dispersal? *Molecular Ecology* 13(3):551-561.

Bjorndal, K.A. 1985. Nutritional ecology of sea turtles. *Copeia* 1985(3):736-751.

Bjorndal, K.A. and G.H. Balazs (eds.). 1983. *Manual of Sea Turtle Research and Conservation Techniques.* Center for Environmental Education, Washington, DC.

Bjorndal, K.A., A.B. Meylan, and B.J. Turner. 1983. Sea turtles nesting at Melbourne Beach, Florida, 1. Size, growth and reproductive biology. *Biological Conservation* 26(1):65-77.

Bjorndal, K.A. and A. Carr. 1989. Variation in clutch size and egg size in the green turtle nesting population at Tortuguero, Costa Rica. *Herpetologica* 45(2):181-189.

Bjorndal, K.A., A.B. Bolten, and C.J. Lagueux. 1994. Ingestion of marine debris by juvenile sea turtles in coastal Florida habitats. *Marine Pollution Bulletin* 28(3):154-158.

Bjorndal, K.A., A.B. Bolten, A.L. Coan, Jr., and P. Kleiber. 1995. Estimation of green turtle (*Chelonia mydas*) growth rates from length-frequency analysis. *Copeia* 1995(1):71-77.

Bjorndal, K.A., J.A. Wetherall, A.B. Bolten, and J.A. Mortimer. 1999. Twenty-six years of green turtle nesting at Tortuguero, Costa Rica: An encouraging trend. *Conservation Biology* 13(1):126-134.

Bjorndal, K.A. and A.B. Bolten (eds.). 2000. *Proceedings of a Workshop on Assessing Abundance and Trends for In-Water Sea Turtle Populations.* NOAA Technical Memorandum NMFS-SEFSC-445, Southeast Fisheries Science Center, National Marine Fisheries Service, National Oceanic and Atmospheric Administration, Miami, Florida.

Bjorndal, K.A., A.B. Bolten, and H.R. Martins. 2000a. Somatic growth model of juvenile loggerhead sea turtles: Duration of the pelagic stage. *Marine Ecology Progress Series* 202:265-272.

Bjorndal, K.A., A.B. Bolten, and M. Chaloupka. 2000b. Green turtle somatic growth model: Evidence for density dependence. *Ecological Applications* 10(1):269-282.

Bjorndal, K.A., A.B. Bolten, B. Koike, B.A. Schroeder, D.J. Shaver, W.G. Teas, and W.N. Witzell. 2001. Somatic growth function for immature loggerhead sea turtles, *Caretta caretta*, in southeastern U.S. waters. *Fishery Bulletin* 99(2):240-246.

Bjorndal, K.A., A.B. Bolten, T. Dellinger, C. Delgado, and H.R. Martins. 2003a. Compensatory growth in oceanic loggerhead sea turtles: Response to a stochastic environment. *Ecology* 84(5):1237-1249.

Bjorndal, K.A., A.B. Bolten, and H.R. Martins. 2003b. Estimates of survival probabilities for oceanic-stage loggerhead sea turtles (*Caretta caretta*) in the North Atlantic. *Fishery Bulletin* 101:732-736.

Bjorndal, K.A., A.B. Bolten, and M. Chaloupka. 2003c. Survival probability estimates for immature green turtles, *Chelonia mydas*, in the Bahamas. *Marine Ecology Progress Series* 252:273-281.

Bjorndal, K.A., A.B. Bolten, and M. Chaloupka. 2005. Evaluating trends in abundance of immature green turtles, *Chelonia mydas*, in the greater Caribbean. *Ecological Applications* 15(1):304-314.

Bjorndal, K.A. and A.B. Bolten. 2008. Annual variation in source contributions to a mixed stock: Implications for quantifying connectivity. *Molecular Ecology* 17(9):2185-2193.

Blanvillain, G., A.P. Pease, A.L. Segars, D.C. Rostal, A.J. Richards, and D.W. Owens. 2008. Comparing methods for the assessment of reproductive activity in adult male loggerhead sea turtles *Caretta caretta* at Cape Canaveral, Florida. *Endangered Species Research* 6:75-85.

Blumenthal, J.M., J.L. Solomon, C.D.L. Bell, T.J. Austin, S.G. Ebanks-Petrie, M.S. Coyne, A.C. Broderick, and B.J. Godley. 2006. Satellite tracking highlights the need for international cooperation in marine turtle management. *Endangered Species Research* 2:51-61.

Blumenthal, J.M., T.J. Austin, C.D.L. Bell, J.B. Bothwell, A.C. Broderick, S.G. Ebanks-Petrie, K.E. Luke, J.R. Olynik, M.F. Orr, J.L. Solomon, and B.J. Godley. 2009a. Ecology of Hawksbill turtles, *Eretmochelys imbricata*, on a western Caribbean foraging ground. *Chelonian Conservation and Biology* 8(1):1-10.

Blumenthal, J.M., F.A. Abreu-Grobis, T.J. Austin, A.C. Broderick, M.W. Bruford, M.S. Coyne, G. Ebanks-Petrie, A. Formia, P.A. Meylan, A.B. Meylan, and B.J. Godley. 2009b. Turtle groups or turtle soup: Dispersal patterns of hawksbill turtles in the Caribbean. *Molecular Ecology* 18(23):4841-4853.

Bolker, B., T. Okuyama, K.A. Bjorndal, and A.B. Bolten. 2003. Sea turtle stock estimation using genetic markers: Accounting for sampling error of rare genotypes. *Ecological Applications* 13(3):763-775.

Bolten, A.B. 2003a. Active swimmers—passive drifters: The oceanic juvenile stage of loggerheads in the Atlantic system. In *Loggerhead Sea Turtles*, Bolten, A.B. and B.E. Witherington (eds.). Smithsonian Books, Washington, DC.

Bolten, A.B. 2003b. Variation in sea turtle life history patterns: Neritic vs. oceanic developmental stages. In *The Biology of Sea Turtles, Volume II*, Lutz, P.L., J.A. Musick, and J. Wyneken (eds.). CRC Press, Boca Raton, Florida.

Bolten, A.B. and K.A. Bjorndal. 1992. Blood profiles for a wild population of green turtles (*Chelonia mydas*) in the southern Bahamas: Size-specific and sex-specific relationships. *Journal of Wildlife Diseases* 28(3):407-413.

Bolten, A.B., J.A. Wetherall, G.H. Balazs, and S.G. Pooley. 1996. *Status of Marine Turtles in the Pacific Ocean Relevant to Incidental Take in the Hawaii-Based Pelagic Longline Fishery.* NOAA Technical Memorandum NMFS-SWFSC-230, Southwest Fisheries Science Center, National Marine Fisheries Service, National Oceanic and Atmospheric Administration, La Jolla, California.

Bolten, A.B., K.A. Bjorndal, H.R. Martins, T. Dellinger, M.J. Biscoito, S.E. Encalada, and B.W. Bowen. 1998. Trans-Atlantic developmental migrations of loggerhead sea turtles demonstrated by mtDNA sequence analyses. *Ecological Applications* 8(1):1-7.

Bolten, A.B., L.B. Crowder, M.G. Dodd, S.L. MacPherson, J.A. Musick, B.A. Schroeder, B.E. Witherington, K.J. Long, and M.L. Snover. In press. Quantifying multiple threats to endangered species: An example from loggerhead sea turtles. *Frontiers in Ecology and the Environment.*

Boulon, R. 2000. Trends in sea turtle strandings, U.S. Virgin Islands: 1982 to 1997. In *Proceedings of the Eighteenth International Sea Turtle Symposium*, Abreu-Grobois, F.A., R. Briseño-Deuñas, R. Márquez-Millán, and L. Sarti-Martínez (eds.). NOAA Technical Memorandum NMFS-SEFSC-436, Southeast Fisheries Science Center, National Marine Fisheries Service, National Oceanic and Atmospheric Administration, Miami, Florida.

Bowen, B.W., A.B. Meylan, and J.C. Avise. 1989. An odyssey of the green sea turtle, *Chelonia mydas*: Ascension Island revisited. *Proceedings of the National Academy of Sciences of the United States of America* 86(2):573-576.

Bowen, B.W., A.B. Meylan, J.P. Ross, C.J. Limpus, G.H. Balazs, and J.C. Avise. 1992. Global population structure and natural history of the green turtle (*Chelonia mydas*) in terms of matriarchal phylogeny. *Evolution* 46(4):865-881.

Bowen, B.W., J.C. Avise, J.I. Richardson, A.B. Meylan, D. Margaritoulis, and S.R. Hopkins-Murphy. 1993. Population structure of the loggerhead turtle (*Caretta caretta*) in the northwest Atlantic Ocean and Mediterranean Sea. *Conservation Biology* 7(4):834-844.

Bowen, B.W., N. Kamezaki, C.J. Limpus, G.R. Hughes, A.B. Meylan, and J.C. Avise. 1994. Global phylogeography of the loggerhead turtle (*Caretta caretta*) as indicated by mitochondrial DNA haplotypes. *Evolution* 48(6):1820-1828.

Bowen, B.W., F.A. Abreu-Grobois, G.H. Balazs, N. Kamezaki, C.J. Limpus, and R.J. Ferl. 1995. Trans-Pacific migrations of the loggerhead sea turtle demonstrated with mitochondrial DNA markers. *Proceedings of the National Academy of Sciences of the United States of America* 92(9):3731-3734.

Bowen, B.W., A.L. Bass, S. Mei-Chow, M. Bostrom, K.A. Bjorndal, A.B. Bolten, T. Okuyama, B.M. Bolker, S.P. Epperly, E. Lacasella, D.J. Shaver, M.G. Dodd, S.R. Hopkins-Murphy, J.A. Musick, M. Swingle, K. Rankin-Baransky, W.G. Teas, W.N. Witzell, and P.H. Dutton. 2004. Natal homing in juvenile loggerhead turtles (*Caretta caretta*). *Molecular Ecology* 13:3797-3808.

Bowen, B.W., A.L. Bass, L. Soares, and R.J. Toonen. 2005. Conservation implications of complex population structure: Lessons from the loggerhead turtle (*Caretta caretta*). *Molecular Ecology* 14(8):2389-2402.

Bowen, B.W. and S.A. Karl. 2007. Population genetics and phylogeography of sea turtles. *Molecular Ecology* 16(23):4886-4907.

Bowen, B.W., W.S. Grant, Z. Hillis-Starr, D.J. Shaver, K.A. Bjorndal, A.B. Bolten, and A.L. Bass. 2007. Mixed-stock analysis reveals the migrations of juvenile hawksbill turtles (*Eretmochelys imbricata*) in the Caribbean Sea. *Molecular Ecology* 16(1):49-60.

Bradshaw, G.A. and J.G. Borchers. 2000. Uncertainty as information: Narrowing the science-policy gap. *Conservation Ecology* 4(1):7.

Braun, J. and S.P. Epperly. 1995. Aerial surveys for sea turtles in southern Georgia waters. In *Proceedings of the Twelfth Annual Workshop on Sea Turtle Biology and Conservation*, Richardson, J.I. and T.H. Richardson (eds.). NOAA Technical Memorandum NMFS-SEFSC-361, Southeast Fisheries Science Center, National Marine Fisheries Service, National Oceanic and Atmospheric Administration, Miami, Florida.

Braun-McNeill, J., C.R. Sasso, and L. Avens. 2007. Estimates of realized survival for juvenile loggerhead sea turtles (*Caretta caretta*) in the United States. *Herpetological Conservation and Biology* 2(2):100-105.

Bresette, M.J., J.C. Gorham, and B. Peery. 1998. Site fidelity and size frequencies of juvenile green turtles (*Chelonia mydas*) utilizing nearshore reefs in St. Lucie County, Florida. *Marine Turtle Newsletter* 82:5-7.

Bresette, M.J., B.E. Witherington, R.M. Herren, D.A. Bagley, J.C. Gorham, S.L. Traxler, C.K. Crady, and R. Hardy. 2010. Size-class partitioning and herding in a foraging group of green turtles *Chelonia mydas*. *Endangered Species Research* 9:105-116.

Brill, R.W. and M.E. Lutcavage. 2001. Understanding environmental influences on movements and depth distributions of tunas and billfishes can significantly improve population assessments. *American Fisheries Society Symposium* 25:179-198.

Broderick, A.C. and B.J. Godley. 1999. Effect of tagging marine turtles on nesting behaviour and reproductive success. *Animal Behaviour* 58(3):587-591.

Broderick, A.C., B.J. Godley, and G.C. Hays. 2001. Trophic status drives interannual variability in nesting numbers of marine turtles. *Proceedings of the Royal Society B: Biological Sciences* 268(1475):1481-1487.

Broderick, A.C., F. Glen, B.J. Godley, and G.C. Hays. 2002. Estimating the number of green and loggerhead turtles nesting annually in the Mediterranean. *Oryx* 36(3):227-235.

Broderick, A.C., F. Glen, B.J. Godley, and G.C. Hays. 2003. Variation in reproductive output of marine turtles. *Journal of Experimental Marine Biology and Ecology* 288(1):95-109.

Brooks, S.P., R. King, and B.J.T. Morgan. 2004. A Bayesian approach to combining animal abundance and demographic data. *Animal Biodiversity and Conservation* 27(1):515-529.

Brooks, S.P., S.N. Freeman, J.J.D. Greenwood, R. King, and C. Mazzetta. 2008. Quantifying conservation concern: Bayesian statistics, birds and the red lists. *Biological Conservation* 141(5):1436-1441.

Buckland, S.T., D.R. Anderson, K.P. Burnham, and J.L. Laake. 1993. *Distance Sampling: Estimating Abundance of Biological Populations*. Chapman and Hall, London, England, United Kingdom.

Buckland, S.T., D.R. Anderson, K.P. Burnham, J.L. Laake, D.L. Borchers, and L. Thomas. 2001. *Introduction to Distance Sampling: Estimating Abundance of Biological Populations*. Oxford University Press, Oxford, England, United Kingdom.

Buckland, S.T., D.R. Anderson, K.P. Burnham, J.L. Laake, D.L. Borchers, and L. Thomas (eds.). 2004. *Advanced Distance Sampling: Estimating Abundance of Biological Populations*. Oxford University Press, New York, New York.

Buitrago, J., H.J. Guada, and E. Doyle. 2008. Conservation science in developing countries: An inside perspective on the struggles in sea turtle research and conservation in Venezuela. *Environmental Science and Policy* 11(6):562-578.

Burger, Jr., L.W. and B.D. Leopold. 2001. Integrating mathematics and statistics into undergraduate wildlife programs. *Wildlife Society Bulletin* 29(4):1024-1030.

Burnham, K.P. 1993. A theory for combined analysis of ring recovery and recapture data. In *Marked Individuals in the Study of Bird Population*, Lebreton, J.D. and P.M. North (eds.). Birkhäuser Verlag, Basel, Switzerland.

Burnham, K.P., D.R. Anderson, and J.L. Laake. 1985. Efficiency and bias in strip and line transect sampling. *Journal of Wildlife Management* 49(4):1012-1018.

Cadrin, S.X., K.D. Friedland, and J.R. Waldman (eds.). 2005. *Stock Identification Methods: Applications in Fisheries Science*. Elsevier Academic Press, San Diego, California.

Caillouet, Jr., C.W. 1998. Testing hypotheses of the Kemp's ridley head-start experiment. *Marine Turtle Newsletter* 79:16-18.

Caillouet, Jr., C.W., D.J. Shaver, W.G. Teas, J.M. Nance, D.B. Revera, and A.C. Cannon. 1996. Relationship between sea turtle stranding rates and shrimp fishing intensities in the northwestern Gulf of Mexico: 1986–1989 versus 1990–1993. *Fishery Bulletin* 94:237-249.

Campbell, C.L. and C.J. Lagueux. 2005. Survival probability estimates for large juvenile and adult green turtles (*Chelonia mydas*) exposed to an artisanal marine turtle fishery in the western Caribbean. *Herpetologica* 61(2):91-103.

Carlsson, J., J.R. McDowell, J.E.L. Carlsson, and J.E. Graves. 2007. Genetic identity of YOY bluefin tuna from the eastern and western Atlantic spawning areas. *Journal of Heredity* 98:23-28.

Carr, A.F. 1963. Panspecific reproductive convergence in *Lepidochelys kempii*. *Ergebnisse der Biologie* 26:298-303.

Case, T.J. 2000. *An Illustrated Guide to Theoretical Ecology*. Oxford University Press, New York, New York.

Castelletti, A. and R. Soncini-Sessa. 2007. Bayesian networks and participatory modeling in water resource management. *Environmental Modelling and Software* 22(8):1075-1088.

Caswell, H. 2001. *Matrix Population Models, Second Edition*. Sinauer and Associates, Sunderland, Massachusetts.

Catchpole, E.A., S.N. Freeman, B.J.T. Morgan, and M.P. Harris. 1998. Integrated recovery/recapture data analysis. *Biometrics* 54(1):33-46.

Caut, S., V. Hulin, and M. Girondot. 2006. Impact of density-dependent nest destruction on emergence success of Guianan leatherback turtles (*Dermochelys coriacea*). *Animal Conservation* 9(2):189-197.

Ceriani, S.A. and J. Wyneken. 2008. Comparative morphology and sex identification of the reproductive system in formalin-preserved sea turtle specimens. *Zoology* 111(3):179-187.

Chaloupka, M. 1998. Polyphasic growth in pelagic loggerhead sea turtles. *Copeia* 1998(2):516-518.

Chaloupka, M. 2001a. System-of-equations growth function for southern Great Barrier Reef green sea turtles. *Chelonian Conservation and Biology* 4:88-93.

Chaloupka, M. 2001b. Historical trends, seasonality and spatial synchrony in green turtle egg production. *Biological Conservation* 101(3):263-279.

Chaloupka, M. 2002a. Stochastic simulation modelling of southern Great Barrier Reef green turtle population dynamics. *Ecological Modelling* 148(1):79-109.

Chaloupka, M. 2002b. *Development of a Stochastic Population Model for the Northwestern Pacific Loggerhead Sea Turtle Stock: Background Material and Model Documentation*. National Marine Fisheries Service, National Oceanic and Atmospheric Administration, Honolulu, Hawaii.

Chaloupka, M. 2003a. Stochastic simulation modelling of loggerhead sea turtle population dynamics given exposure to competing mortality risks in the western south Pacific. In *Loggerhead Sea Turtles*, Bolten, A.B. and B.E. Witherington (eds.). Smithsonian Books, Washington, DC.

Chaloupka, M. 2003b. *Phase 2—Development of a Population Simulation Model for the Southern Great Barrier Reef Green Turtle Stock*. Great Barrier Reef Marine Park Authority Research Publication No. 81, Great Barrier Reef Marine Park Authority, Townsville, Queensland, Australia.

Chaloupka, M. 2004. Exploring the metapopulation dynamics of the southern Great Barrier Reef green turtle stock and possible consequences of sex-biased local harvesting. In *Species Conservation and Management: Case Studies*, Akçakaya, H., M. Burgman, O. Kindvall, C. Wood, P. Sjogren-Gulve, J. Hattfield, and M. McCarthy (eds.). Oxford University Press, New York, New York.

Chaloupka, M. 2007. Using Bayesian belief networks to evaluate the relative risk of fishing gear types on sea turtle stocks. In *Report of the Workshop on Assessing the Relative Importance of Sea Turtle Mortality Due to Fisheries in Southeast Asia*. Food and Agriculture Organization of the United Nations, Rome, Italy.

Chaloupka, M. and J.A. Musick. 1997. Age, growth, and population dynamics. In *The Biology of Sea Turtles, Volume II*, Lutz, P.L. and J.A. Musick (eds.). CRC Press, Boca Raton, Florida.

Chaloupka, M. and C.J. Limpus. 2001. Trends in the abundance of sea turtles resident in southern Great Barrier Reef waters. *Biological Conservation* 102(3):235-249.

Chaloupka, M. and C.J. Limpus. 2002. Survival probability estimates for the endangered loggerhead sea turtle resident in southern Great Barrier Reef waters. *Marine Biology* 140:267-277.

Chaloupka, M., D. Parker, and G.H. Balazs. 2004a. Modeling post-release mortality of pelagic loggerhead sea turtles exposed to the Hawaii-based pelagic longline fishery. *Marine Ecology Progress Series* 280:285-293.

Chaloupka, M., C.J. Limpus, and J.D. Miller. 2004b. Green turtle somatic growth dynamics in a spatially disjunct Great Barrier Reef metapopulation. *Coral Reefs* 23(3):325-335.

Chaloupka, M. and G.H. Balazs. 2005. Modelling the effect of fibropapilloma disease on the somatic growth dynamics of Hawaiian green sea turtles. *Marine Biology* 147(5):1251-1260.

Chaloupka, M. and C.J. Limpus. 2005. Estimates of sex- and age-class-specific survival probabilities for a southern Great Barrier Reef green sea turtle population. *Marine Biology* 146(6):1251-1261.

Chaloupka, M. and G.H. Balazs. 2007. Using Bayesian state-space modeling to assess the recovery and harvest potential of the Hawaiian green turtle stock. *Ecological Modeling* 205(1-2):93-109.

Chaloupka, M., N. Kamezaki, and C.J. Limpus. 2008a. Is climate change affecting the population dynamics of the endangered Pacific loggerhead sea turtle? *Journal of Experimental Marine Biology and Ecology* 356(1-2):136-143.

Chaloupka, M., T.M. Work, G.H. Balazs, S.K.K. Murakawa, and R. Morris. 2008b. Cause-specific temporal and spatial trends in green sea turtle strandings in the Hawaiian Archipelago (1982–2003). *Marine Biology* 154(5):887-898.

Chaloupka, M., K.A. Bjorndal, G.H. Balazs, A.B. Bolten, L. Ehrhart, C.J. Limpus, H. Suganuma, S. Troëng, and M. Yamaguchi. 2008c. Encouraging outlook for recovery of a once severely exploited marine megaherbivore. *Global Ecology and Biogeography* 17(2):297-304.

Chan, E.H. 1989. White spot development, incubation and hatching success of leatherback turtle (*Dermochelys coriacea*) eggs from Rantau Abang, Malaysia. *Copeia* 1989(1):42-47.

Chan, E.H. and H.C. Liew. 1996. Decline of the leatherback population in Terengganu, Malaysia, 1956–1995. *Chelonian Conservation and Biology* 2(2):196-203.

Cheng, I.J. and T.H. Chen. 1997. The incidental capture of five species of sea turtles by coastal setnet fisheries in the eastern waters of Taiwan. *Biological Conservation* 82(2):235-239.

Clark, T.W. 2001. Developing policy-oriented curricula for conservation biology: Professional and leadership education in the public interest. *Conservation Biology* 15(1):31-39.

Clarke, J.T. and S.A. Norman. 2005. Results and evaluation of U.S. Navy shock trial environmental mitigation of marine mammals and sea turtles. *Journal of Cetacean Research and Management* 7(1):43-50.

Conant, T.A., P.H. Dutton, T. Eguchi, S.P. Epperly, C.C. Fahy, M.H. Godfrey, S.L. MacPherson, E.E. Possardt, B.A. Schroeder, J.A. Seminoff, M.L. Snover, C.M. Upite, and B.E. Witherington. 2009. *Loggerhead Sea Turtle (*Caretta caretta*) 2009 Status Review under the U.S. Endangered Species Act*. Report of the Loggerhead Biological Review Team to the National Marine Fisheries Service.

Cooney, R. 2004. *The Precautionary Principle in Biodiversity Conservation and Natural Resource Management: An Issues Paper for Policy-Makers, Researchers and Practitioners*. IUCN Policy and Global Change Series No. 2, International Union for the Conservation of Nature, Gland, Switzerland.

Cotter, A.J.R., L. Burt, C.G.M. Paxton, C. Fernandez, S.T. Buckland, and J.X. Pan. 2004. Are stock assessment methods too complicated? *Fish and Fisheries* 5(3):235-254.

Cotter, J., P. Petitgas, A. Abella, P. Apostolaki, B. Mesnil, C.Y. Politou, J. Rivoirard, M.J. Rochet, M.T. Spedicato, V.M. Trenkel, and M. Woillez. 2009. Towards an ecosystem approach to fisheries management (EAFM) when trawl surveys provide the main source of information. *Aquatic Living Resources* 22(2):243-254.

Crawford, N.G. 2009. SMOGD: Software for the measurement of genetic diversity. *Molecular Ecology Resources* 10310(3):556-557.

Crim, J.L., L.D. Spotila, J.R. Spotila, M. O'Connor, R. Reina, C.J. Williams, and F.V. Paladino. 2002. The leatherback turtle, *Dermochelys coriacea*, exhibits both polyandry and polygyny. *Molecular Ecology* 11(10):2097-2106.

Crowder, L.B., D.T. Crouse, S.S. Heppell, and T.H. Martin. 1994. Predicting the impact of turtle excluder devices on loggerhead sea turtle populations. *Ecological Applications* 4(3):437-445.

Crowder, L.B., S.R. Hopkins-Murphy, and J.A. Royle. 1995. Effects of turtle excluder devices (TEDs) on loggerhead sea turtle strandings with implications for conservation. *Copeia* 1995:773-779.

Dambacher, J.M., H.W. Li, and P.A. Rossignol. 2003. Qualitative predictions in model ecosystems. *Ecological Modelling* 161(1-2):79-93.

Dambacher, J.M., D.J. Gaughan, M.J. Rochet, P.A. Rossigniol, and V.M. Trenkel. 2009. Qualitative modelling and indicators of exploited ecosystems. *Fish and Fisheries* 10(3):305-322.

Daniels, R.C., T.W. White, and K. Chapman. 2006. Sea-level rise: Destruction of threatened and endangered species habitat in South Carolina. *Journal of Environmental Management* 17(3):373-385.

Dennis, B., P.L. Munholland, and J.M. Scott. 1991. Estimation of growth and extinction parameters for endangered species. *Ecological Monographs* 61(2):115-143.

Dethmers, K.E.M., D. Broderick, C. Moritz, N.N. FitzSimmons, C.J. Limpus, S. Lavery, S. Whiting, M. Guinea, R.I.T. Prince, and R. Kennet. 2006. The genetic structure of Australasian green turtles (*Chelonia mydas*): Exploring the geographical scale of genetic exchange. *Molecular Ecology* 15(13):3931-3946.

Diez, C.E. and R.P. van Dam. 2002. Habitat effect on hawksbill turtle growth rates on feeding grounds at Mona and Monito Islands, Puerto Rico. *Marine Ecology Progress Series* 234:301-309.

Dinno, A. 2007. Loop analysis of causal feedback in epidemiology: Illustration relating to urban neighborhoods and resident depressive experiences. *Social Science and Medicine* 65(10): 2043-2057.

Dodge, K.D., R. Prescott, D. Lewis, D. Murley, and C. Merigo. 2007. A review of cold stun strandings on Cape Cod, Massachusetts from 1979–2003. In *Proceedings of the Twenty-Fourth Annual Symposium on Sea Turtle Biology and Conservation*, Mast, R.B., B.J. Hutchinson, and A.H. Hutchinson (eds.). NOAA Technical Memorandum NMFS-SEFSC-567, Southeast Fisheries Science Center, National Marine Fisheries Service, National Oceanic and Atmospheric Administration, Miami, Florida.

Drummond, A.J. and A. Rambaut. 2007. BEAST: Bayesian evolutionary analysis by sampling trees. *BMC Evolutionary Biology* 7:214.

Duke University. 2009. *OBIS SEAMAP*. [Online]. Available: http://seamap.env.duke.edu/ [2010, May 21].

Dupanloup, I., S. Schneider, and L. Excoffier. 2002. A simulated annealing approach to define the genetic structure of populations. *Molecular Ecology* 11(12):2571-2581.

Dutton, D.L., P.H. Dutton, M. Chaloupka, and R.H. Boulon. 2005. Increase of a Caribbean leatherback turtle *Dermochelys coriacea* nesting population linked to long-term nest protection. *Biological Conservation* 126(2):186-194.

Dutton, P.H., B.W. Bowen, D.W. Owens, A. Barragan, and S.K. Davis. 1999. Global phylogeography of the leatherback turtle, *Dermochelys coriacea*. *Journal of Zoology* 248:397-409.

Dutton, P.H., C. Hitipeuw, M. Zein, S.R. Benson, G. Petro, J. Pita, V. Rei, L. Ambio, and J. Bakarbessy. 2007. Status and genetic structure of nesting stocks of leatherback turtles (*Dermochelys coriacea*) in the western Pacific. *Chelonian Conservation and Biology* 6(1):47-53.

Dutton, P.H., G.H. Balazs, R.A. LeRoux, S.K.K. Murakawa, P. Zarate, and L.S. Martinez. 2008. Composition of Hawaiian green turtle foraging aggregations: mtDNA evidence for a distinct regional population. *Endangered Species Research* 5:37-44.

Dutton, P.H. and A. Frey. 2009. Characterization of polymorphic microsatellite markers for the green turtle (*Chelonia mydas*). *Molecular Ecology Resources* 9(1):354-356.

Eaton, C., E. McMichael, B.E. Witherington, A. Foley, R. Hardy, and A.B. Meylan. 2008. *In-Water Sea Turtle Monitoring and Research in Florida: Review and Recommendations*. NOAA Technical Memorandum NMFS-OPR-38, National Marine Fisheries Service, National Oceanic and Atmospheric Administration, Silver Spring, Maryland.

Eckert, S.A. 2002. Distribution of juvenile leatherback sea turtle *Dermochelys coriacea* sightings. *Marine Ecology Progress Series* 230:289-293.

Eckert, S.A. and H.R. Martins. 1989. Transatlantic travel by juvenile loggerhead turtle. *Marine Turtle Newsletter* 45:15.

Eckert, K.L., K.A. Bjorndal, F.A. Abreu-Grobois, and M. Donnelly (eds.). 1999. *Research and Management Techniques for the Conservation of Sea Turtles*. IUCN/SSC Marine Turtle Specialist Group Publication No. 4, International Union for the Conservation of Nature, Gland, Switzerland.

Eguchi, T. and T. Gerrodette. 2009. A Bayesian approach to line-transect analysis for estimating abundance. *Ecological Modelling* 220(13-14):1620-1630.

Ehrhart, L.M. and L.H. Ogren. 1999. Studies in foraging habitats: Capturing and handling turtles. In *Research and Management Techniques for the Conservation of Sea Turtles*, Eckert, K.L., K.A. Bjorndal, F.A. Abreu-Grobois, and M. Donnelly (eds.). IUCN/SSC Marine Turtle Specialist Group Publication No. 4, International Union for the Conservation of Nature, Gland, Switzerland.

Ehrhart, L.M., D.A. Bagley, and W.E. Redfoot. 2003. Loggerhead turtles in the Atlantic Ocean: Geographic distribution, abundance, and population status. In *Loggerhead Sea Turtles*, Bolten, A.B. and B.E. Witherington (eds.). Smithsonian Books, Washington, DC.

Encalada, S.E., P.N. Lahanas, K.A. Bjorndal, A.B. Bolten, M.M. Miyamoto, and B.W. Bowen. 1996. Phylogeography and population structure of the green turtle (*Chelonia mydas*) in the Atlantic Ocean and Mediterranean Sea: A mitochondrial DNA control region sequence assessment. *Molecular Ecology* 5(4):473-484.

Encalada, S.E., K.A. Bjorndal, A.B. Bolten, J.C. Zurita, B.A. Schroeder, E. Possardt, C.J. Sears, and B.W. Bowen. 1998. Population structure of loggerhead turtle (*Caretta caretta*) nesting colonies in the Atlantic and Mediterranean regions as inferred from mtDNA control region sequences. *Marine Biology* 130(4):567-575.

Engeman, R.M., R.E. Martin, H.T. Smith, J. Woolard, C.K. Crady, S.A. Shwiff, B. Constatin, M. Stahl, and J. Griner. 2005. Dramatic reduction in predation on marine turtle nests through improved predator monitoring and management. *Oryx* 39(3):318-326.

Engstrom, T.N., T. Edwards, M. Ostentowski, and E. Meyers. 2007. A compendium of PCR primers for mtDNA, microsatellite and other nuclear markers for freshwater turtles and tortoises. *Chelonian Research Monographs* 4:124-141.

Epperly, S.P., J. Braun, and A. Veishlow. 1995a. Sea turtles in North Carolina waters. *Conservation Biology* 9(2):384-394.

Epperly, S.P., J. Braun, and A.J. Chester. 1995b. Aerial surveys for sea turtles in North Carolina inshore waters. *Fishery Bulletin* 93:254-261.

Epperly, S.P., J. Braun, A.J. Chester, F.A. Cross, J.V. Merriner, P.A. Tester, and J.H. Churchill. 1996. Beach strandings as an indicator of at-sea mortality of sea turtles. *Bulletin of Marine Science* 59(2):289-297.

Epperly, S.P. and W.G. Teas. 2002. Turtle excluder devices—Are the escape openings large enough? *Fishery Bulletin* 100:466-474

Epperly, S.P., J. Braun-McNeill, and P.M. Richards. 2007. Trends in catch rates of sea turtles in North Carolina, USA. *Endangered Species Research* 3:283-293.

Epstein, M. 1999. Incidental impact to nesting Wilson's plovers during the sea turtle nest monitoring season. *Florida Field Naturalist* 27(4):173-176.

Excoffier, L., P.E. Smouse, and J.M. Quattro. 1992. Analysis of molecular variance inferred from metric distances among DNA haplotypes: Application to human mitochondrial DNA restriction data. *Genetics* 131:479-491.

Excoffier, L., G. Laval, and S. Schneider. 2005. Arlequin (version 3.0): An integrated software package for population genetics data analysis. *Evolutionary Bioinformatics Online* 1:47-50.

Fish, M.R., I.M. Cote, J.A. Gill, A.P. Jones, S. Renshoff, and A.R. Watkinson. 2005. Predicting the impact of sea-level rise on Caribbean sea turtle nesting habitat. *Conservation Biology* 19:482-491.

FitzSimmons, N.N., C. Moritz, and S.S. Moore. 1995. Conservation and dynamics of microsatellite loci over 300 million years of marine turtle evolution. *Molecular Biology and Evolution* 12:432-440.

FitzSimmons, N.N., C.J. Limpus, J.A. Norman, A.R. Goldizen, J.D. Miller, and C. Moritz. 1997a. Philopatry of male marine turtles inferred from mitochondrial DNA markers. *Proceedings of the National Academy of Sciences of the United States of America* 94(16):8912-8917.

FitzSimmons, N.N., C. Moritz, C.J. Limpus, L. Pope, and R. Prince. 1997b. Geographic structure of mitochondrial and nuclear gene polymorphisms in Australian green turtle populations and male-biased gene flow. *Genetics* 147(4):1843-1854.

Florida Fish and Wildlife Conservation Commission. 2007. *Marine Turtle Conservation Guidelines*. Florida Fish and Wildlife Conservation Commission, Tallahassee, Florida.

Florida Fish and Wildlife Conservation Commission. 2009. *Florida Statewide Nesting Beach Survey Data—2009 Season*. [Online]. Available: http://research.myfwc.com/features/view_article.asp?id=11812 [2010, May 3].

Fonnesbeck, C.J. and M.J. Conroy. 2004. Application of integrated Bayesian modeling and Markov chain Monte Carlo methods to the conservation of a harvested species. *Animal Biodiversity and Conservation* 27(1):267-281.

Fossette, S., P. Gaspar, Y. Handrich, J.Y. Georges, and Y. Le Maho. 2008. Dive and beak movement patterns in leatherback turtles *Dermochelys coriacea* during internesting intervals in French Guiana. *Journal of Animal Ecology* 77(2):236-246.

Frazer, N.B. 1983. Survivorship of adult female loggerhead sea turtles, *Caretta caretta*, nesting on Little Cumberland Island, Georgia, USA. *Herpetologica* 39(4):436-447.

Frazer, N.B. 1987. Preliminary estimates of survivorship for wild juvenile loggerhead sea turtles (*Caretta caretta*). *Journal of Herpetology* 21(3):232-235.

Frazer, N.B. and J.I. Richardson. 1985. Seasonal variation in clutch size for loggerhead sea turtles, *Caretta caretta*, nesting on Little Cumberland Island, Georgia, USA. *Copeia* 1985:1083-1085.

Frick, J. 1976. Orientation and behaviour of hatchling green turtles (*Chelonia mydas*) in the sea. *Animal Behaviour* 24(4):849-857.

Fritts, T.H., W. Hoffman, and M.A. McGehee. 1983. The distribution and abundance of marine turtles in the Gulf of Mexico and nearby Atlantic waters. *Journal of Herpetology* 17(4):327-344.

Fulton, E.A., A.D.M. Smith, and C.R. Johnson. 2003. Effect of complexity on marine ecosystem models. *Marine Ecology Progress Series* 253:1-16.

Fulton, E.A., A.D.M. Smith, and A.E. Punt. 2005. Which ecological indicators can robustly detect effects of fishing? *ICES Journal of Marine Science* 62(3):540-551.

Gardner, S. and W. Nichols. 2001. Assessment of sea turtle mortality rates in the Bahia Magdalena region, Baja California Sur, Mexico. *Chelonian Conservation and Biology* 4:197-199.

Gates, C.E., R.A. Valverde, C.L. Mo, A.C. Chaves, J. Ballesteros, and J. Peskin. 1996. Estimating arribada size using a modified instantaneous count procedure. *Journal of Agricultural, Biological, and Environmental Statistics* 1(3):275-287.

Genome 10K Community of Scientists. 2009. Genome 10K: A proposal to obtain whole-genome sequence for 10,000 vertebrate species. *Journal of Heredity* 100(6):659-674.

Gerrodette, T. 2000. Estimating abundance with transects. In *Proceedings of a Workshop on Assessing Abundance and Trends for In-Water Sea Turtle Populations*, Bjorndal, K.A. and A.B. Bolten (eds.). NOAA Technical Memorandum NMFS-SEFSC-445, Southeast Fisheries Science Center, National Marine Fisheries Service, National Oceanic and Atmospheric Administration, Miami, Florida.

Gilman, E., D. Kobayashi, T. Swenarton, N. Brothers, P. Dalzell, and I. Kinan-Kelly. 2007. Reducing sea turtle interactions in the Hawaii-based longline swordfish fishery. *Biological Conservation* 139(1-2):19-28.

Gilman, E., N. King, T. Peterson, V. Chavan, and A. Hahn. 2009. Building the biodiversity data commons—The global biodiversity information facility. In *ICT for Agriculture and Biodiversity Conservation*, Maurer, L. (ed.). ICT Ensure, Graz University of Technology, Graz, Austria.

Girondot, M., A.D. Tucker, P. Rivalan, M.H. Godfrey, and J. Chevalier. 2002. Density-dependent nest destruction and population fluctuations of Guianan leatherback turtles. *Animal Conservation* 5(1):75-84.

Godley, B.J., A.C. Broderick, and G.C. Hays. 2001. Nesting of green turtles (*Chelonia mydas*) at Ascension Island, South Atlantic. *Biological Conservation* 97(2):151-158.

Grant, W.S., G.B. Milner, P. Krasnowski, and F.M. Utter. 1980. Use of biochemical genetic variants for identification of sockeye salmon (*Oncorhynchus nerka*) stocks in Cook Inlet, Alaska. *Canadian Journal of Fisheries and Aquatic Sciences* 37(8):1236-1247.

Groves, R.M., R.B. Cialdini, and M.P. Couper. 1992. Understanding the decision to participate in a survey. *The Public Opinion Quarterly* 56(4):475-495.

Hamann, M., C.J. Limpus, and D.W. Owens. 2003. Reproductive cycles of males and females. In *The Biology of Sea Turtles, Volume II*, Lutz, P.L., J.A. Musick, and J. Wyneken (eds.). CRC Press, Boca Raton, Florida

Hanski, I. 1999. *Metapopulation Ecology*. Oxford University Press, Oxford, England, United Kingdom.

Hanski, I. and M.E. Gilpin. 1991. Metapopulation dynamics: Brief history and conceptual domain. *Biological Journal of the Linnean Society* 42(1-2):3-16.

Hanski, I. and D. Simberloff. 1997. The metapopulation approach, its history, conceptual domain, and application to conservation. In *Metapopulation Biology: Ecology, Genetics, and Evolution*, Hanski, I. and M.E. Gilpin (eds.). Academic Press, San Diego, California.

Hard, J.J. 1995. Science, education, and the fisheries scientist. *Fisheries* 20(3):10-16.

Harrison, A.L. and K.A. Bjorndal. 2006. Connectivity and wide-ranging species in the ocean. In *Connectivity Conservation*, Crooks, K.R. and M.A. Sanjayan (eds.). Cambridge University Press, Cambridge, Massachusetts.

Hart, K., P. Mooreside, and L.B. Crowder. 2006. Interpreting the spatio-temporal patterns of sea turtle strandings: Going with the flow. *Biological Conservation* 129(2):283-290.

Hartl, D.L. and A.G. Clark. 2007. *Principles of Population Genetics, Fourth Edition*. Sinauer Associates, Sunderland, Massachusetts.

Hatase, H., N. Takai, Y. Matsuzawa, W. Sakamoto, K. Omuta, K. Goto, N. Arai, and T. Fujiwara. 2002. Size-related differences in feeding habitat use of adult female loggerhead turtles *Caretta caretta* around Japan determined by stable isotope analyses and satellite telemetry. *Marine Ecology Progress Series* 233:272-281.

Hawkes, L.A., A.C. Broderick, M.S. Coyne, M.H. Godfrey, L.F. Lopez-Suarez, S.E. Merino, N. Varo-Cruz, and B.J. Godley. 2006. Phenotypically linked dictomy in sea turtle foraging requires multiple conservation approaches. *Current Biology* 16(10):990-995.

Hawkes, L.A., A.C. Broderick, M.H. Godfrey, and B.J. Godley. 2007. Investigating the potential impacts of climate change on a marine turtle population. *Global Change Biology* 13(5):923-932.

Hays, G.C. 2000. The implications of variable remigration intervals for the assessment of population size in marine turtles. *Journal of Theoretical Biology* 206:221-227.

Hays, G.C. and R. Marsh. 1997. Estimating the age of juvenile loggerhead sea turtles in the North Atlantic. *Canadian Journal of Zoology* 75:40-46.

Hedges, M.E. 2007. *Development and Application of a Multistate Model to the Northern Subpopulation of Loggerhead Sea Turtles (*Caretta caretta*).* Master of Science Thesis, Virginia Polytechnic Institute and State University, Blacksburg, Virginia.

Heithaus, M.R., J.J. McLash, A. Frid, L.M. Dill, and G.J. Marshall. 2002. Novel insights into green sea turtle behaviour using animal-borne video cameras. *Journal of the Marine Biological Association of the United Kingdom* 82:1049-1050.

Henwood, T. and S.P. Epperly. 1999. Aerial surveys in foraging habitats. In *Research and Management Techniques for the Conservation of Sea Turtles*, Eckert, K.L., K.A. Bjorndal, F.A. Abreu-Grobois, and M. Donnelly (eds.). IUCN/SSC Marine Turtle Specialist Group Publication No. 4, International Union for the Conservation of Nature, Gland, Switzerland.

Heppell, S.S., H. Caswell, and L.B. Crowder. 2000. Life histories and elasticity patterns: Perturbation analysis for species with minimal demographic data. *Ecology* 81(3):654-665.

Heppell, S.S., M.L. Snover, and L.B. Crowder. 2002. Sea turtle population dynamics. In *Biology of Sea Turtles, Volume II,* Lutz, P., J. Wyneken, and J.A. Musick (eds.). CRC Press, Boca Raton, Florida.

Heppell, S.S., L.B. Crowder, D.T. Crouse, S.P. Epperly, and N.B. Frazer. 2003. Population models for the Atlantic loggerhead. In *Loggerhead Sea Turtles*, Bolten, A.B. and B.E. Witherington (eds.). Smithsonian Books, Washington, DC.

Heppell, S.S., D.T. Crouse, L.B. Crowder, S.P. Epperly, W. Gabriel, T. Henwood, R. Marquez, and N.B. Thompson. 2005. A population model to estimate recovery time, population size, and management impacts on Kemp's ridley sea turtles. *Chelonian Conservation and Biology* 4(4):767-773.

Heppell, S.S., P. Burchfield, and J. Peña. 2007. Kemp's ridley recovery: How far have we come, and where are we headed? In *Biology and Conservation of Ridley Sea Turtles,* Plotkin, P. (ed.). Johns Hopkins University Press, Baltimore, Maryland.

Hey, J. 2010a. Isolation with migration models for more than two populations. *Molecular Biology and Evolution* 27(4):905-920.

Hey, J. 2010b. *Hey Lab Distributed Software.* [Online]. Available: http://genfaculty.rutgers.edu/hey/software [2010, May 12].

Hey, J. and R. Nielsen. 2007. Integration within the Felsenstein equation for improved Markov chain Monte Carlo methods in population genetics. *Proceedings of the National Academy of Sciences of the United States of America* 104(8):2785-2790.

Higgins, B.M., B.A. Robertson, and T.D. Williams. 1997. *Manual for Mass Wire Tagging of Hatchling Sea Turtles and the Detection of Internal Wire Tags.* NOAA Technical Memorandum NMFS-SEFC-402, National Marine Fisheries Service, National Oceanic and Atmospheric Administration, Silver Spring, Maryland.

Hilborn, R. 1985. Apparent stock recruitment relationships in mixed stock fisheries. *Canadian Journal of Fisheries and Aquatic Sciences* 42(4):718-723.

Hilborn, R. and C.J. Walters. 1992. *Quantitative Fisheries Stock Assessment: Choice, Dynamics and Uncertainty.* Chapman and Hall, New York, New York.

Hilborn, R. and M. Mangel. 1997. *The Ecological Detective: Confronting Models with Data.* Princeton University Press, Princeton, New Jersey.

Hildebrand, H. 1963. Hallazgo del área de anidación de la tortuga "lora" *Lepidochelys kempii* (Garman), en la costa Occidental del Golfo de México (Rept., Chel.). *Ciencia (México)* 22(4):105-112.

Hochscheid, S., F. Maffucci, F. Bentivegna, and R.P. Wilson. 2005. Gulps, wheezes, and sniffs: How measurement of beak movement in sea turtles can elucidate their behaviour and ecology. *Journal of Experimental Marine Biology and Ecology* 316(1):45-53.

Holmes, E.E. 2001. Estimating risks in declining populations with poor data. *Proceedings of the National Academy of Sciences of the United States of America* 98(9):5072-5077.

Holmes, E.E. 2004. Beyond theory to application and evaluation: Diffusion approximations for population viability analysis. *Ecological Applications* 14(4):1272-1293.

Hopkins-Murphy, S.R., T.M. Murphy, C.P. Hope, J.W. Coker, and M.E. Hoyle. 2001. *Population Trends and Nesting Distribution of the Loggerhead Turtle (*Caretta caretta*) in South Carolina 1980-1997.* Final Report to the U.S. Fish and Wildlife Service, Wildlife Diversity Section, Division of Wildlife and Freshwater Fisheries, South Carolina Department of Natural Resources, Columbia, South Carolina.

Horikoshi, K. 1992. *Egg Survivorship and Primary Sex Ratio of Green Turtles,* Chelonia mydas, *at Tortuguero, Costa Rica.* Unpublished Ph.D. Dissertation, University of Florida, Gainesville, Florida.

Hosack, G., K. Hayes, and J.M. Dambacher. 2008. Assessing model structure uncertainty through an analysis of system feedback and Bayesian networks. *Ecological Applications* 18(4):1070-1082.

Hubisz, M.J., D. Falush, M. Stephens, and J.K. Pritchard. 2009. Inferring weak population structure with the assistance of sample group information. *Molecular Ecology* Resources 9(5):1322-1332.

Hulot, F.D., G. Lacroix, F. Lescher-Moutoué, and M. Loreau. 2000. Functional diversity governs ecosystem response to nutrient enrichment. *Nature* 405(6784):340-344.

Hunter, J.R. and J.H. Churnside. 1995. *Airborne Fishery Assessment Technology: A NOAA Workshop Report.* Administration Report LJ-95-02, Southwest Fisheries Science Center, National Marine Fisheries Service, National Oceanic and Atmospheric Administration, La Jolla, California.

International Union for Conservation of Nature. 2010. *The IUCN Red List of Threatened Species.* [Online]. Available: http://www.iucnredlist.org [2010, April 29].

Jackson, A.L., A.C. Broderick, W.J. Fuller, F. Glen, G.D. Ruxton, and B.J. Godley. 2008. Sampling design and its effect on population monitoring: How much monitoring do turtles really need? *Biological Conservation* 141:2932-2941.

Jackson, J.B.C., M.X. Kirby, W.H. Berger, K.A. Bjorndal, L.W. Botsford, B.J. Bourque, R.H. Bradbury, R. Cooke, J. Erlandson, J.A. Estes, T.P. Hughes, S. Kidwell, C.B. Lange, H.S. Lenihan, J.M. Pandolfi, C.H. Peterson, R.S. Steneck, M.J. Tegner, and R.R. Warner. 2001. Historical overfishing and the recent collapse of coastal ecosystems. *Science* 293(5530):629-638.

Jacobson, S.K. 1990. Graduate education in conservation biology. *Conservation Biology* 4(4):431-440.

Jacobson, S.K. and J.G. Robinson. 1990. Training the new conservationist: Cross-disciplinary education in the 1990s. *Environmental Conservation* 17(4):319-327.

James, M.C. and N. Mrosovsky. 2004. Body temperatures of leatherback turtles (*Dermochelys coriacea*) in temperate waters off Nova Scotia, Canada. *Canadian Journal of Zoology* 82(8):1302-1306.

James, M.C., S. Sherrill-Mix, K. Martin, and R.A. Myers. 2006. Canadian waters provide critical foraging habitat for leatherback sea turtles. *Biological Conservation* 133(3):347-357.

Jeffreys, A.J. 2005. Genetic fingerprinting. *Nature Medicine* 11(10):35-39.

Jennings, S. 2005. Indicators to support an ecosystem approach to fisheries. *Fish and Fisheries* 6(3):212-232.

Jensen, M.P., F.A. Abreu-Grobois, J. Frydenberg, and V. Loeschcke. 2006. Microsatellites provide insight into contrasting mating patterns in arribada vs. non-arribada olive ridley sea turtle rookeries. *Molecular Ecology* 15(9):2567-2575.

Jones, C.M. 2006. Estuarine and diadromous fish metapopulations. In *Marine Metapopulations*, Krtizer, J.P. and P.F. Sale (eds.). Elsevier Academic Press, Burlington, Massachusetts.

Jost, L. 2008. G_{ST} and its relatives do not measure differentiation. *Molecular Ecology* 17(18):4015-4026.

Julian, F. and M. Beeson. 1998. Estimates of marine mammal, turtle and seabird mortality for two California gillnet fisheries: 1990–1995. *Fishery Bulletin* 96:271-284.

Kamezaki, N. and M. Matsui. 1997. A review of biological studies on sea turtles in Japan. *Japan Journal of Herpetology* 17(1):16-32.

Kamezaki, N., Y. Matsuzawa, K. Omuta, O. Abe, K. Goto, I. Wakabayashi, H. Takeshita, Y. Nakashima, J. Kodama, M. Ishii, Y. Nakamura, T. Iwamoto, M. Samejima, T. Kamata, K. Mizobuchi, H. Kato, M. Hayami, S. Hagino, T. Tanaka, A. Yamamoto, T. Toji, H. Suganuma, T. Fujii, H. Asakawa, M. Uematsu, A. Naruse, T. Yamato, and Y. Kondo. 2003. Population change of Japanese loggerhead turtles, *Caretta caretta*, estimated by nesting number. In *The Biology and Conservation of Loggerhead Sea Turtles*, Bolten, A.B. and B.E. Witherington (eds.). University of Florida Press, Gainesville, Florida.

Karl, S.A., B.W. Bowen, and J.C. Avise. 1992. Global population structure and male-mediated gene flow in the green turtle (*Chelonia mydas*): RFLP analysis of anonymous nuclear DNA regions. *Genetics* 131:163-173.

Karl, S.A. and J.C. Avise. 1993. PCR-based assays of Mendelian polymorphisms from anonymous single-copy nuclear DNA: Techniques and applications for population genetics. *Molecular Biology and Evolution* 10:342-361.

Kendall, W.L. and R. Bjorkland. 2001. Using open robust design models to estimate temporary emigration from capture-recapture data. *Biometrics* 57(4):1113-1122.

Kendall, W.L. and J.D. Nichols. 2002. Estimating state-transition probabilities for unobservable states using capture–recapture/resighting data. *Ecology* 83(12):3276-3284.

Kenney, R.D. and C.R. Shoop. In press. Aerial surveys of marine turtles. In *Reptile Biodiversity: Standard Methods for Inventory and Monitoring*, McDiarmid, R.W., M.S. Foster, C. Guyer, J.W. Gibbons, and N. Chernoff (eds.). University of California Press, Berkeley, California.

Kichler, K., M.T. Holder, S.K. Davis, S.R. Marquez, and D.W. Owens. 1999. Detection of multiple paternity in the Kemp's ridley sea turtle with limited sampling. *Molecular Ecology* 8(5):819-830.

Kimura, M. and G.H. Weiss. 1964. The stepping stone model of population structure and the decrease of genetic correlation with distance. *Genetics* 49(4):561-576.

Kingman, J.F.C. 1982. The coalescent. *Stochastic Processes and Their Applications* 13(3):235-248.

Kritzer, J.P. and P.F. Sale (eds.). 2006. *Marine Metapopulations*. Academic Press, New York, New York.

Kroll, A.J. 2007. Integrating professional skills in wildlife student education. *The Journal of Wildlife Management* 71(1):226-230.

Kubis, S., M. Chaloupka, L. Ehrhart, and M.J. Bresette. 2009. Growth rates of juvenile green turtles *Chelonia mydas* from three ecologically distinct foraging habitats along the east central coast of Florida, USA. *Marine Ecology Progress Series* 389:257-269.

Kudo, H., A. Murakami, and S. Watanabe. 2003. Effects of sand hardness and human beach use on emergence success of loggerhead sea turtles on Yakushima Island, Japan. *Chelonian Conservation and Biology* 4:695-696.

Kuhner, M.K. 2009. Coalescent genealogy samplers: Windows into population history. *Trends in Ecology and Evolution* 24(2):86-93.

Laurent, L., P. Casale, M.N. Bradai, B.J. Godley, G. Gerosa, A.C. Broderick, W. Schroth, B. Schierwater, A.M. Levy, D. Freggi, E.M. Abd el-Mawla, D.A. Hadoud, H.E. Gomati, M. Domingo, M. Hadjichristophorou, L. Kornarky, F. Demirayak, and C. Gautier. 1998. Molecular resolution of the marine turtle stock composition in fishery bycatch: A case study in the Mediterranean. *Molecular Ecology* 7(11):1529-1542.

Lauret-Stepler, M., J. Bourjea, D. Roos, D. Pelletier, P.G. Ryan, S. Ciccione, and H. Grizel. 2007. Reproductive seasonality and trend of *Chelonia mydas* in the SW Indian Ocean: A 20 yr study based on track counts. *Endangered Species Research* 3:217-227.

Le Gall, J.Y., P. Bosc, D. Chateau, and M. Taquet. 1986. An estimation of the number of adult females of green turtles *Chelonia mydas* per nesting season at Tromelin and Europa (Indian Ocean) (1973–1985). *Oceanographique Tropicale* 21(1):3-22.

Le Tissier, M.D.A., J.M. Hills, J.A. McGregor, and M. Ireland. 2004. A training framework for understanding conflict in the coastal zone. *Coastal Management* 22(1):77-88.

LeBlanc, A.M. and T. Wibbels. 2009. Effect of daily water treatment on hatchling sex ratios in a turtle with temperature-dependent sex determination. *Journal of Experimental Zoology* 311A(1):68-72.

LeBuff, Jr., C.R. 1974. Unusual nesting relocation in the loggerhead turtle *Caretta caretta*. *Herpetologica* 30(1):29-31.

Lee, P.L.M. 2008. Molecular ecology of marine turtles: New approaches and future directions. *Journal of Experimental Marine Biology and Ecology* 356(1-2):25-42.

Lee, P.L.M., P. Luschi, and G.C. Hays. 2007. Detecting female precise natal philopatry in green turtles using assignment methods. *Molecular Ecology* 16(1):61-74.

Lehodey, P., I. Senina, and R. Murtugudde. 2008. A spatial ecosystem and populations dynamics model (SEAPODYM)—Modeling of tuna and tuna-like populations. *Progress in Oceanography* 78(4):304-318.

Leon, Y.M. and C.E. Diez. 1999. Population structure of hawksbill turtles on a foraging ground in the Dominican Republic. *Chelonian Conservation Biology* 3(2):230-236.

Levins, R. 1966. The strategy of model building in population biology. In *Conceptual Issues in Evolutionary Biology, First Edition*, Sober, E. (ed.). MIT Press, Cambridge, Massachusetts.

Levins, R. 1969. Some demographic and genetic consequences of environmental heterogeneity for biological control. *Bulletin of the Entomological Society of America* 15:237-240.

Levins, R. 1970. Extinction. In *Some Mathematical Problems in Biology*, Gerstenhaber, M. (ed.). American Mathematical Society, Providence, Rhode Island.

Lewison, R.L., L.B. Crowder, and D.J. Shaver. 2003. The impact of turtle excluder devices and fisheries closures on loggerhead and Kemp's ridley strandings in the western Gulf of Mexico. *Conservation Biology* 17(4):1089-1097.

Lewison, R.L., S.N. Freeman, and L.B. Crowder. 2004. Quantifying the effects of fisheries on threatened species: The impact of pelagic longlines on loggerhead and leatherback sea turtles. *Ecology Letters* 7(3):221-231.

Liermann, K. and R. Hilborn. 2001. Depensation: Evidence, models, and implications. *Fish and Fisheries* 2(1):33-58.

Limpus, C.J. 1992. Observations on first breeding by a loggerhead turtle. *Marine Turtle Newsletter* 56:1-2.

Limpus, C.J. 1994. The loggerhead turtle, *Caretta caretta*, in Queensland: Feeding ground selection following her first nesting season. In *Proceedings of the Fourteenth Annual Symposium on Sea Turtle Biology and Conservation*, Bjorndal, K.A., A.B. Bolten, D.A. Johnson, and P.J. Eliazar (eds.). NOAA Technical Memorandum NMFS-SEFSC-351, Southeast Fisheries Science Center, National Marine Fisheries Service, National Oceanic and Atmospheric Administration, Miami, Florida.

Limpus, C.J. 2009. *A Biological Review of Australian Marine Turtles*. Queensland Environmental Protection Agency, Brisbane, Queensland, Australia.

Limpus, C.J., P.C. Reed, and J.D. Miller. 1983. Islands and turtles: The influence of choice of nesting beach on sex ratio. In *Proceedings of the Inaugural Great Barrier Reef Conference*, Baker, J., R. Carter, P. Sammarco, and K. Stark (eds.). James Cook University Press, Townville, Queensland, Australia.

Limpus, C.J. and P.C. Reed. 1985a. Green sea turtles stranded by Cyclone Kathy on the south-western coast of the Gulf of Carpentaria. *Australian Wildlife Research* 12(3):523-533.

Limpus, C.J. and P.C. Reed. 1985b. The green turtle, *Chelonia mydas*, in Queensland: A preliminary description of the population structure in a coral reef feeding ground. In *Biology of Australasian Frogs and Reptiles*, Grigg, G., R. Shine, and H. Ehmann (eds.). Royal Society of New South Wales, Sydney, Australia.

Limpus, C.J., P.C. Reed, and J.D. Miller. 1985. Temperature dependent sex determination in Queensland sea turtles: Intraspecific variation in *Caretta caretta*. In *Biology of Australasian Frogs and Reptiles*, Grigg, G., R. Shine, and H. Ehmann (eds.). Royal Society of New South Wales, Sydney, New South Wales, Australia.

Limpus, C.J., J.D. Miller, C.J. Parmenter, D. Reimer, N. McLachlan, and R. Webb. 1992. Migration of green (*Chelonia mydas*) and loggerhead (*Caretta caretta*) turtles to and from eastern Australian rookeries. *Wildlife Research* 19(3):347-358.

Limpus, C.J., P.J. Couper, and M.A. Read. 1994. The loggerhead turtle, *Caretta caretta*, in Queensland: Population structure in a warm temperate feeding area. *Memoirs of the Queensland Museum* 37:195-204.

Limpus, C.J. and N. Nicholls. 2000. ENSO regulation of Indo-Pacific green turtle populations. In *The Australian Experience*, Hammer, G.L., N. Nicholls, and C. Mitchell (eds.). Kluwer Academic Publishers, Dordrecht, The Netherlands.

Limpus, C.J. and D.J. Limpus. 2001. The loggerhead turtle, *Caretta caretta*, in Queensland: Breeding migrations and fidelity to a warm temperate feeding area. *Chelonian Conservation and Biology* 4:142-153.

Limpus, C.J. and D.J. Limpus. 2003a. The biology of the loggerhead turtle, *Caretta caretta*, in southwest Pacific Ocean foraging areas. In *The Biology and Conservation of Loggerhead Sea Turtles*, Bolten, A.B. and B.E. Witherington (eds.). University of Florida Press, Gainesville, Florida.

Limpus, C.J. and D.J. Limpus. 2003b. Loggerhead turtles in the equatorial and southern Pacific Ocean: A species in decline. In *Loggerhead Sea Turtles*, Bolten, A.B. and B.E. Witherington (eds.). Smithsonian Books, Washington, DC.

Limpus, C.J., J.D. Miller, C.J. Parmenter, and D.J. Limpus. 2003. The green turtle, *Chelonia mydas*, population of Raine Island and the northern Great Barrier Reef: 1843–2001. *Memoirs of the Queensland Museum* 49:349-440.

Limpus, C.J., D.J. Limpus, M. Munchow, and P. Barnes. 2005. *Queensland Turtle Conservation Project: Raine Island Turtle Study, 2004–2005*, Jones, M. (ed.). Freshwater and Marine Sciences Unit, Environmental Sciences Division, Environmental Protection Agency, Queensland Parks and Wildlife Service, Queensland, Australia.

Lin, G., A. Chang, H.W. Yap, and G.H. Yue. 2008. Characterization and cross-species amplification of microsatellites from the endangered Hawksbill turtle (*Eretmochelys imbricate*). *Conservation Genetics* 9(4):1071-1073.

Loiselle, S., G.M. Carpento, V. Hull, T. Waller, and C. Rossi. 2000. Feedback analysis in reserve management: Studying local myths using qualitative models. *Ecological Modelling* 129(1):25-37.

López-Castro, M.C. and A. Rocha-Olivares. 2005. The panmixia paradigm of the eastern Pacific olive ridley turtles revised: Consequences for their conservation and evolutionary biology. *Molecular Ecology* 14(11):3325-3334.

Lutcavage, M.E. and P.L. Lutz. 1997. Diving physiology. In *The Biology of Sea Turtles, Volume I*, Lutz, P.L. and J.A. Musick (eds.). CRC Press, Boca Raton, Florida.

Lutcavage, M.E., P. Plotkin, B.E. Witherington, and P.L. Lutz. 1997. Human impacts on sea turtle survival. In *The Biology of Sea Turtles, Volume I*, Lutz, P.L. and J.A. Musick (eds.). CRC Press, Boca Raton, Florida.

Maier, P.P., A.L. Segars, M.D. Arendt, J.D. Whitaker, B.W. Stender, L.P. Parker, R. Vendetti, D.W. Owens, J. Quattro, and S.R. Murphy. 2004. *Development of an Index of Sea Turtle Abundance Based Upon In-Water Sampling with Trawl Gear*. South Carolina Department of Natural Resources, Charleston, South Carolina.

Makowski, C., R. Slattery, and M. Salmon. 2005. "Shark fishing": A method for determining the abundance and distribution of sea turtles at shallow reef habitats. *Herpetological Review* 36(1):36-38.

Manni, F., E. Guérard, and E. Heyer. 2004. Geographic patterns of (genetic, morphologic, linguistic) variation: How barriers can be detected by Monmonier's algorithm. *Human Biology* 76(2):173-190.

Mansfield, K.L. 2006. *Sources of Mortality, Movements and Behavior of Sea Turtles in Virginia*. Ph.D. Dissertation, College of William and Mary, Williamsburg, Virginia.

Mansfield, K.L., V.S. Saba, J.A. Keinath, and J.A. Musick. 2009a. Satellite tracking reveals a dichotomy in migration strategies among juvenile loggerhead turtles in the Northwest Atlantic. *Marine Biology* 156(12):2555-2570.

Mansfield, K.L., J. Wyneken, and D. Rittschof. 2009b. *Technological Development and Adaptation of Small-Scale Satellite Tags to Characterize Long-Term Dispersal Patterns and Early Life History of Neonate Sea Turtles: Semi-Annual Progress Report*. [Online]. Available: http://www.largepelagics.unh.edu/pdf/Progress%20Reports/LPRC_progress%20report_MANSFIELD_Oct%2009.pdf [2010, April 29].

Manski, C.E. 1993. Identification problems in the social sciences. In *Sociological Methodology*, Marsden, P. (ed.). Blackwell, Oxford, England, United Kingdom.

Marcovaldi, M.A., M.H. Godfrey, and N. Mrosovsky. 1997. Estimating sex ratios of loggerhead turtles in Brazil from pivotal incubation durations. *Canadian Journal of Zoology* 75:755-770.

Marcovaldi, M.A. and G.G. Marcovaldi. 1999. Marine turtles of Brazil: The history and structure of Projeto TAMAR-IBAMA. *Biological Conservation* 91(1):35-41.

Marcovaldi, M.A. and M. Chaloupka. 2007. Conservation status of the loggerhead sea turtle in Brazil: An encouraging outlook. *Endangered Species Research* 3:133-143.

Marsh, H. and D.F. Sinclair. 1989. An experimental evaluation of dugong and sea turtle aerial survey techniques. *Australian Wildlife Research* 16(6):639-650.

Marsh, H. and W.K. Saalfeld. 1989. A survey of sea turtles in the northern Great Barrier Reef Marine Park. *Australian Wildlife Research* 16:239-249.

Massey, W.E. 1989. Science education in the United States: What the scientific community can do. *Science* 245(4921):915-921.

Matsuzawa, Y., K. Sato, W. Sakamoto, and K.A. Bjorndal. 2002. Seasonal fluctuations in sand temperature: Effects on the incubation period and mortality of loggerhead sea turtle (*Caretta caretta*) pre-emergent hatchlings in Minabe, Japan. *Marine Biology* 140(3):639-646.

Matter, W.J. and R.J. Steidl. 2000. University undergraduate curricula in wildlife beyond 2000. *Wildlife Society Bulletin* 28(3):503-507.

Maunder, M.N. 2003. Paradigm shifts in fisheries stock assessment: From integrated analysis to Bayesian analysis and back again. *Natural Resource Modeling* 16(4):465-475.

Maunder, M.N. 2004. Population viability analysis based on combining Bayesian, integrated, and hierarchical analyses. *Acta Oecologica* 26(2):85-94.

Mazaris, A.D., Ø. Fiksen, and Y.G. Matsinos. 2005. Using an individual-based model for assessment of sea turtle population viability. *Population Ecology* 47(3):179-191.

Mazaris, A.D., B. Broder, and Y.G. Matsinos. 2006. An individual based model of a sea turtle population to analyze effects of age dependent mortality. *Ecological Modelling* 198(1-2):174-182.

Mazaris, A.D., Y.G. Matsinos, and J. Pantis. 2009. Evaluating the impacts of coastal squeeze on sea turtle nesting. *Ocean and Coastal Management* 52(2):139-145.

McClellan, C.M. and A.J. Read. 2007. Complexity and variation in loggerhead sea turtle life history. *Biology Letters* 3(6):592-594.

McClellan, C.M., J. Braun-McNeill, L. Avens, B.P. Wallace, and A.J. Read. 2010. Stable isotopes confirm a foraging dichotomy in juvenile loggerhead sea turtles. *Journal of Experimental Marine Biology and Ecology* 387(1-2):44-51.

McClenachan, L., J.B.C. Jackson, and M.J.H. Newman. 2006. Conservation implications of historic sea turtle nesting beach loss. *Frontiers in Ecology and the Environment* 4(6):290-296.

McLachlan, N., B. McLachlan, J. McLachlan, and B. McLachlan. 2006. *Queensland Turtle Conservation Project, Wreck Rock Study 2005–2006*, Jones, M. (ed.). Conservation Technical and Data Report Volume 6 ISSN 1449–194X, Freshwater and Marine Sciences Unit, Environmental Sciences Division, Environmental Protection Agency, Queensland Parks and Wildlife Service, Queensland, Australia.

Merkhofer, M.W. 1987. *Decision Science and Social Risk Management: Technology, Risk, and Society*. Springer, Dordrecht, The Netherlands.

Merrick, R. and H. Haas. 2008. *Analysis of Atlantic Sea Scallop (*Placopecten magellanicus*) Fishery Impacts on the North Atlantic Population of Loggerhead Sea Turtle*s (Caretta caretta*)*. NOAA Technical Memorandum NMFS-NE-207, Northeast Fisheries Science Center, National Marine Fisheries Service, National Oceanic and Atmospheric Administration, Woods Hole, Massachusetts.

Meylan, A.B. 1999. Status of the hawksbill turtle (*Eretmochelys imbricata*) in the Caribbean region. *Chelonian Conservation and Biology* 3(2):177-184.

Miller, J.D. 1997. Reproduction in sea turtles. In *The Biology of Sea Turtles, Volume I*, Lutz, P.L. and J.A. Musick (eds.). CRC Press, Boca Raton, Florida.

Miller, J.D. 1999. Determining clutch size and hatching success. In *Research and Management Techniques for the Conservation of Sea Turtles*, Eckert, K.L., K.A. Bjorndal, F.A. Abreu-Grobois, and M. Donnelly (eds.). IUCN/SSC Marine Turtle Specialist Group Publication No. 4, International Union for the Conservation of Nature, Gland, Switzerland.

Miro-Herrans, A.T., X. Velez-Zuazo, J.P. Acevedo, and W.O. McMillan. 2008. Isolation and characterization of novel microsatellites from the critically endangered hawksbill sea turtle (*Eretmochelys imbricata*). *Molecular Ecology Resources* 8(5):1098-1101.

Monzón-Argüello, C., J. Muñoz, A. Marco, L.F. López-Jurado, and C. Rico. 2008. Twelve new polymorphic microsatellite markers from the loggerhead sea turtle (*Caretta caretta*) and cross-species amplification on other marine turtle species. *Conservation Genetics* 9(4):1045-1049.

Moritz, C., S. Lavery, and R. Slade. 1995. Using allele frequency and phylogeny to define units for conservation and management. In *Evolution and the Aquatic Ecosystem: Defining Unique Units in Population Conservation*, Nielsen, J.L. (ed.). American Fisheries Society Symposium 17, American Fisheries Society, Bethesda, Maryland.

Mrosovsky, N. and M.H. Godfrey. 1995. Manipulating sex ratios: Turtle speed ahead! *Chelonian Conservation and Biology* 1:238-240.

Mrosovsky, N., C. Lavin, and M.H. Godfrey. 1995. Thermal effects of condominiums on a turtle beach in Florida. *Biological Conservation* 74(3):151-156.

Murray, K.T. 2008. *Estimated Average Annual Bycatch of Loggerhead Sea Turtles (*Caretta caretta*) in U.S. Mid-Atlantic Bottom Otter Trawl Gear, 1996–2004 (Second Edition)*. Northeast Fisheries Science Center, National Marine Fisheries Service, National Oceanic and Atmospheric Administration, Woods Hole, Massachusetts.

Murray, K.T. 2009. *Proration of Estimated Bycatch of Loggerhead Sea Turtles in U.S. Mid-Atlantic Sink Gillnet Gear to Vessel Trip Report Landed Catch, 2002–2006*. Northeast Fisheries Science Center, National Marine Fisheries Service, National Oceanic and Atmospheric Administration, Woods Hole, Massachusetts.

Musick, J.A., D.E. Barnard, and J.A. Keinath. 1994. Aerial estimates of seasonal distribution and abundance of sea turtles near the Cape Hatteras faunal barrier. In *Proceedings of the Thirteenth Annual Symposium on Sea Turtle Biology and Conservation*, Schroeder, B.A. and B.E. Witherington (eds.). NOAA Technical Memorandum NMFS-SEFSC-341, Southeast Fisheries Science Center, National Marine Fisheries Service, National Oceanic and Atmospheric Administration, Miami, Florida.

Myers, A.E. and G.C. Hays. 2006. Do leatherback turtles forage during the breeding season? A combination of data-logging devices provide new insights. *Marine Ecology Progress Series* 322:259-267.

National Marine Fisheries Service. 1991. *Dredging of Channels in the Southeastern United States from North Carolina through Cape Canaveral, Florida: Biological Opinion Dated November 25, 1991*. Southeast Regional Office, National Marine Fisheries Service, National Oceanic and Atmospheric Administration, St. Petersburg, Florida.

National Marine Fisheries Service. 2005. *Endangered Species Act Section 7 Consultation Biological Opinion*. [Online]. Available: http://www.nero.noaa.gov/nero/hotnews/seaturtles/ScallopDec04BO.pdf [2010, May 7].

National Marine Fisheries Service and U.S. Fish and Wildlife Service. 2007a. *Green Sea Turtle (*Chelonia mydas*) 5-Year Review: Summary and Evaluation*. U.S. Department of Commerce and U.S. Department of the Interior, Washington, DC.

National Marine Fisheries Service and U.S. Fish and Wildlife Service. 2007b. *Loggerhead Sea Turtle (*Caretta caretta*) 5-Year Review: Summary and Evaluation*. U.S. Department of Commerce and U.S. Department of the Interior, Washington, DC.

National Marine Fisheries Service and U.S. Fish and Wildlife Service. 2007c. *Leatherback Sea Turtle (*Dermochelys coriacea*) 5-Year Review: Summary and Evaluation*. U.S. Department of Commerce and U.S. Department of the Interior, Washington, DC.

National Marine Fisheries Service and U.S. Fish and Wildlife Service. 2007d. *Hawksbill Sea Turtle (*Eretmochelys imbricata*) 5-Year Review: Summary and Evaluation*. U.S. Department of Commerce and U.S. Department of the Interior, Washington, DC.

National Marine Fisheries Service and U.S. Fish and Wildlife Service. 2007e. *Olive Ridley Sea Turtle (*Lepidochelys olivacea*) 5-Year Review: Summary and Evaluation*. U.S. Department of Commerce and U.S. Department of the Interior, Washington, DC.

National Marine Fisheries Service and U.S. Fish and Wildlife Service. 2007f. *Kemp's Ridley Sea Turtle (*Lepidochelys kempii*) 5-Year Review: Summary and Evaluation*. U.S. Department of Commerce and U.S. Department of the Interior, Washington, DC.

National Marine Fisheries Service and U.S. Fish and Wildlife Service. 2008. *Recovery Plan for the Northwest Atlantic Population of the Loggerhead Sea Turtle (*Caretta caretta*), Second Revision*. U.S. Department of Commerce and U.S. Department of the Interior, Washington, DC.

National Marine Fisheries Service Southeast Fisheries Science Center. 2001. *Stock Assessments of Loggerhead and Leatherback Sea Turtles and an Assessment of the Impact of the Pelagic Longline Fishery on the Loggerhead and Leatherback Sea Turtles of the Western North Atlantic*. NOAA Technical Memorandum NMFSSEFSC-455, Southeast Fisheries Science Center, National Marine Fisheries Service, National Oceanic and Atmospheric Administration, Miami, Florida.

National Marine Fisheries Service Southeast Fisheries Science Center. 2008. *Sea Turtle Research Techniques Manual*. NOAA Technical Memorandum NMFS-SEFSC 579, Southeast Fisheries Science Center, National Marine Fisheries Service, National Oceanic and Atmospheric Administration, Miami, Florida.

National Marine Fisheries Service Southeast Fisheries Science Center. 2009. *An Assessment of Loggerhead Sea Turtles to Estimate Impacts of Mortality Reductions on Population Dynamics*. Southeast Fisheries Science Center Contribution PRD-08/09-14, Southeast Fisheries Science Center, National Marine Fisheries Service, National Oceanic and Atmospheric Administration, Miami, Florida.

National Research Council. 1990. *Decline of the Sea Turtles: Causes and Prevention*. National Academy Press, Washington, DC.

National Research Council. 1998. *Improving Fish Stock Assessments*. National Academy Press, Washington, DC.

National Research Council. 2000. *Recruiting Fisheries Scientists: Workshop on Stock Assessment and Social Science Careers*. National Academy Press, Washington, DC.

National Research Council. 2004. *Improving the Use of the "Best Scientific Information Available" Standard in Fisheries Management*. National Academies Press, Washington, DC.

National Research Council. 2009. *Ensuring the Integrity, Accessibility, and Stewardship of Research Data in the Digital Age*. National Academies Press, Washington, DC.

Nielsen, J.L., H. Arrizabalaga, N. Fragoso, A. Hobday, M.E. Lutcavage, and J. Sibert (eds.). 2009. *Tagging and Tracking of Marine Animals with Electronic Devices. Reviews: Methods and Technologies in Fish Biology and Fisheries Volume 9*. Springer, Dordrecht, The Netherlands.

Norman, J.A., C. Moritz, and C.J. Limpus. 1994. Mitochondrial DNA control region polymorphisms: Genetic markers for ecological studies of marine turtles. *Molecular Ecology* 3(4):363-373.

O'Dor, R.K. and M.J.W. Stokesbury. 2009. The ocean tracking network—Adding marine animal movements to the global ocean observing system. In *Tagging and Tracking of Marine Animals with Electronic Devices. Reviews: Methods and Technologies in Fish Biology and Fisheries Volume 9*, Nielsen, J.L., H. Arrizabalaga, N. Fragoso, A. Hobday, M.E. Lutcavage, and J. Sibert (eds.). Springer, Dordrecht, The Netherlands.

Okuyama, T. and B.M. Bolker. 2005. Combining genetic and ecological data to estimate sea turtle origins. *Ecological Applications* 15(1):315-325.

Paetkau, D., W. Calvert, I. Stirling, and C. Strobeck. 1995. Microsatellite analysis of population structure in Canadian polar bears. *Molecular Ecology* 4(3):347-354.

Palka, D. 2000. Abundance and distribution of sea turtles estimated from data collected during cetacean surveys. In *Proceedings of a Workshop on Assessing Abundance and Trends for In-Water Sea Turtle Populations*, Bjorndal, K.A. and A.B. Bolten (eds.). NOAA Technical Memorandum NMFS-SEFSC-445, Southeast Fisheries Science Center, National Marine Fisheries Service, National Oceanic and Atmospheric Administration, Miami, Florida.

Peare, T. and P.G. Parker. 1996. Local genetic structure within two rookeries of *Chelonia mydas* (the green turtle). *Heredity* 77(6):619-628.

Peckham, S.H., D.D. Maldonado, A. Walli, G. Ruiz, L.B. Crowder, and W.J. Nichols. 2007. Small-scale fisheries bycatch jeopardizes endangered Pacific loggerhead turtles. *PLoS ONE* 2(10):e1041.

Pine, W.E., K.H. Pollock, J.E. Hightower, T.J. Kwak, and J.A. Rice. 2003. A review of tagging methods for estimating fish population size and components of mortality. *Fisheries* 28(10):10-23.

Plagányi, É.E. 2007. *Models for an Ecosystem Approach to Fisheries*. Food and Agriculture Organization of the United Nations, Rome, Italy.

Poiner, I.R. and A.N.M. Harris. 1996. The incidental capture, direct mortality and delayed mortality of turtles in Australia's northern prawn fishery. *Marine Biology* 125(4):813-825.

Poloczanska, E.S., C.J. Limpus, and G.C. Hays. 2009. Vulnerability of marine turtles to climate change. In *Advances in Marine Biology, Volume 56*, Sims, D.W. (ed.). Elsevier Academic Press, Burlington, Massachusetts.

Polovina, J.J., D.R. Kobayashi, D.M. Parker, M.P. Seki, and G.H. Balazs. 2000. Turtles on the edge: Movement of loggerhead turtles (*Caretta caretta*) along oceanic fronts, spanning longline fishing grounds in the central North Pacific, 1997–1998. *Fisheries Oceanography* 9(1):71-82.

Pradel, R., J.E. Hines, J.D. Lebreton, and J.D. Nichols. 1997. Capture-recapture survival models taking account of transients. *Biometrics* 53:60-72.

Pritchard, J.K., M. Stephens, and P. Donnelly. 2000. Inference of population structure using multilocus genotype data. *Genetics* 155:945-959.

Puccia, C.J. and R. Levins. 1985. *Qualitative Modeling of Complex Systems: An Introduction to Loop Analysis and Time Averaging*. Harvard University Press, Cambridge, Massachusetts.

Reich, K.J., K.A. Bjorndal, and A.B. Bolten. 2007. The 'lost years' of green turtles: Using stable isotopes to study cryptic lifestages. *Biology Letters* 3(6):712-714.

Reina, R.D., K.J. Abernathy, G.J. Marshall, and J.R. Spotila. 2005. Respiratory frequency, dive behavior and social interactions of leatherback turtles, *Dermochelys coriacea* during the internesting interval. *Journal of Experimental Marine Biology and Ecology* 316(1):1-16.

Rice, J.C. and M.J. Rochet. 2005. A framework for selecting a suite of indicators for fisheries management. *ICES Journal of Marine Science* 62(3):516-527.

Richardson, J.I., D.B. Hall, P.A. Mason, K.M. Andrews, R. Bjorkland, Y. Cai, and R. Bell. 2006. Eighteen years of saturation tagging data reveal a significant increase in nesting hawksbill sea turtles (*Eretmochelys imbricata*) on Long Island, Antigua. *Animal Conservation* 9:302-307.

Richardson, T.H., J.I. Richardson, C. Ruckdeschel, and M.W. Dix. 1978. Remigration patterns of loggerhead sea turtles (*Caretta caretta*) nesting on Little Cumberland and Cumberland Island, Georgia. *Florida Marine Research* 33:39-44.

Rivalan, P., M.H. Godfrey, A.C. Prevot-Julliard, and M. Girondot. 2005a. Maximum likelihood estimates of tag loss in leatherback sea turtles. *Journal of Wildlife Management* 69:540-548.

Rivalan, P., A.C. Prevot-Julliard, R. Choquet, R. Pradel, B. Jacquemin, and M. Girondot. 2005b. Trade-off between current reproductive effort and delay to next reproduction in the leatherback sea turtle. *Oecologia* 145(4):564-574.

Rivalan, P., P.H. Dutton, E. Baudry, S.E. Roden, and M. Girondot. 2006a. Demographic scenario inferred from genetic data in leatherback turtles nesting in French Guiana and Suriname. *Biological Conservation* 130(1):1-9.

Rivalan, P., R. Pradel, R. Choquet, M. Girondot, and A.C. Prevot-Julliard. 2006b. Estimating clutch frequency in the sea turtle *Dermochelys coriacea* using stopover duration. *Marine Ecology Progress Series* 317:285-295.

Roark, A.M., K.A. Bjorndal, and A.B. Bolten. 2009a. Compensatory responses to food restriction in juvenile green turtles (*Chelonia mydas*). *Ecology* 90(9):2524-2534.

Roark, A.M., C. Leeuwenburgh, K.A. Bjorndal, and A.B. Bolten. 2009b. Biochemical indices as correlates of recent growth in juvenile green turtles (*Chelonia mydas*). *Journal of Experimental Marine Biology and Ecology* 376(2):59-67.

Roberts, M.A., T.S. Schwartz, and S.A. Karl. 2004. Global population structure and male-mediated gene flow in the green sea turtle (*Chelonia mydas*): Analysis of microsatellite loci. *Genetics* 166:1857-1870.

Rockwood, L.L. 2006. *Introduction to Population Ecology*. Blackwell Publishing, Malden, Massachusetts.

Saba, V.S., P. Santidrian-Tomillo, R.D. Reina, J.R. Spotila, J.A. Musick, D.A. Evans, and F.V. Paladino. 2007. The effect of the El Niño Southern Oscillation on the reproductive frequency of eastern Pacific leatherback turtles. *Journal of Applied Ecology* 44(2):395-404.

Saba, V.S., G.L. Shillinger, A.M. Swithenbank, B.A. Block, J.R. Spotila, J.A. Musick, and F.V. Paladino. 2008. An oceanographic context for the foraging ecology of eastern Pacific leatherback turtles: Consequences of ENSO. *Deep-Sea Research Part I: Oceanographic Research Papers* 55(5):646-660.

Saladin, N. 2007. *Using Surveys as a Tool to Assess Sea Turtle Distribution and Habitat Use, and to Promote Public Engagement in Endangered Species Research and Management*. Unpublished Master of Science Thesis, Duke University, Durham, North Carolina.

Salmon, M., M.G. Tolbert, D.P. Painter, M. Goff, and R. Reiners. 1995. Behavior of loggerhead sea turtles on an urban beach. II. Hatchling orientation. *Journal of Herpetology* 29(4):568-576.

Santora, C. 2003. Management of turtle bycatch: Can endangered species be protected while minimizing socioeconomic impacts? *Coastal Management* 31(4):423-434.

Sasso, C.R., J. Braun-McNeill, L. Avens, and S.P. Epperly. 2006. Effects of transients on estimating survival and population growth in juvenile loggerhead turtles. *Marine Ecology Progress Series* 324:287-292.

Sasso, C.R. and S.P. Epperly. 2007. Survival of pelagic juvenile loggerhead turtles in the open ocean. *Journal of Wildlife Management* 71:1830-1835.

Schroeder, B.A. and N.B. Thompson. 1987. Distribution of the loggerhead turtle, *Caretta caretta*, and the leatherback turtle, *Dermochelys coreacea*, in the Cape Canaveral, Florida areas: Results of aerial surveys. In *Ecology of East Florida Sea Turtles: Proceedings of the Cape Canaveral, Florida Sea Turtle Workshop, Miami, Florida, February 26-27, 1985*, Witzell, W.N. (ed.). NOAA Technical Report NMFS 53, Southeast Fisheries Center, National Marine Fisheries Service, National Oceanic and Atmospheric Administration, Miami, Florida.

Schroeder, B.A. and S. Murphy. 1999. Population surveys (ground and aerial) on nesting beaches. In *Research and Management Techniques for the Conservation of Sea Turtles*, Eckert, K.L., K.A. Bjorndal, F.A. Abreu-Grobois, and M. Donnelly (eds.). IUCN/SSC Marine Turtle Specialist Group Publication No. 4, International Union for the Conservation of Nature, Gland, Switzerland.

Schroth, W., B. Streit, and B. Schierwater. 1996. Evolutionary handicap for turtles. *Nature* 384:521-522.

Seaturtle.org. 2009. *Satellite Tracking*. [Online]. Available: http://www.seaturtle.org/tracking/ [2010, May 3].

Seaturtle.org. 2010a. *Sea Turtle Nest Monitoring System: STNMS Overview*. [Online]. Available: http://www.seaturtle.org/nestdb/ [2010, May 3].

Seaturtle.org. 2010b. *Sea Turtle Rehabilitation and Necropsy Database*. [Online]. Available: http://www.seaturtle.org/strand/ [2010, May 3].

Selkoe, K.A. and R.J. Toonen. 2006. Microsatellites for ecologists: A practical guide to using and evaluating microsatellite markers. *Ecology Letters* 9(5):615-629.

Seminoff, J.A (ed.). 2002. *2002 IUCN Red List Global Status Assessment, Green Turtle (*Chelonia mydas*), Marine Turtle Specialist Group Review*. Marine Turtle Specialist Group, World Conservation Union, Species Survival Commission, Red List Programme, Gland, Switzerland.

Seminoff, J.A., T.T. Jones, A. Resendiz, W.J. Nichols, and M. Chaloupka. 2003. Monitoring green turtles (*Chelonia mydas*) at a coastal foraging area in Baja California, Mexico: Using multiple indices to describe population status. *Journal of the Marine Biological Association of the United Kingdom* 83(6):1355-1362.

Seminoff, J.A., T.T. Jones, and G.J. Marshall. 2006. Underwater behaviour of green turtles monitored with video-time-depth recorders: What's missing from dive profiles? *Marine Ecology Progress Series* 322:269-280.

Shamblin, B.M., B.C. Faircloth, M.G. Dodd, A. Wood-Jones, S.B. Castleberry, J.P. Carroll, and C.J. Nairn. 2007. Tetranucleotide microsatellites from the loggerhead sea turtle (*Caretta caretta*). *Molecular Ecology Notes* 7(5):784-787.

Shamblin, B.M., B.C. Faircloth, M.G. Dodd, D.A. Bagley, L.M. Ehrhart, P.H. Dutton, A. Frey, and C.J. Nairn. 2009. Tetranucleotide markers from the loggerhead sea turtle (*Caretta caretta*) and their cross-amplification in other marine turtle species. *Conservation Genetics* 10(3):577-580.

Shaver, D.J. and W.G. Teas. 1999. Stranding and salvage networks. In *Research and Management Techniques for the Conservation of Sea Turtles*, Eckert, K.L., K.A. Bjorndal, F.A. Abreu-Grobois, and M. Donnelly (eds.). IUCN/SSC Marine Turtle Specialist Group Publication No. 4, International Union for the Conservation of Nature, Gland, Switzerland.

Shoop, C.R. and R.D. Kenney. 1992. Seasonal distributions and abundances of loggerhead and leatherback sea turtles in waters of the northeastern United States. *Herpetological Monographs* 6:43-67.

Shoop, C.R., C.A. Ruckdeschel, and R.D. Kenney. 1999. Long-term trends in size of stranded juvenile loggerhead sea turtles (*Caretta caretta*). *Chelonian Conservation and Biology* 3:501-504.

Sims, M., R. Bjorkland, P. Mason, and L.B. Crowder. 2008. Statistical power and sea turtle nesting beach surveys: How long and when? *Biological Conservation* 141(12):2921-2931.

Slatkin, M. 1995. A measure of population subdivision based on microsatellite allele frequencies. *Genetics* 139(1):457-462.

Slay, C.K., M. Zani, C. Emmons, E. LaBreque, B. Pike, S.D. Kraus, and R.D. Kenney. 2002. *Early Warning System 1994–2002: Aerial Surveys to Reduce Ship/Whale Collisions in the North Atlantic Right Whale Calving Ground*. Final Report 2002 Edition. Prepared by Edgerton Research Laboratory, New England Aquarium, Boston, Massachusetts for U.S. Department of Commerce, National Ocean Service, Center for Coastal Environmental Health and Biomolecular Research, Charleston, South Carolina.

Smith, A.D.M., K.J. Sainsbury, and R.A. Stevens. 1999. Implementing effective fisheries-management systems—Management strategy evaluation and the Australian partnership approach. *ICES Journal of Marine Science* 56(6):967-979.

Smith, M.H., H.O. Hillstad, M.N. Manlove, D.O. Straney, and J.M. Dean. 1978. Management implications of genetic variability in loggerhead and green sea turtles. In *Proceedings of the 13th International Congress of Game Biologists*, Smith, M.H. (ed.). Savannah River Ecology Laboratory, Aiken, South Carolina.

Snover, M.L. 2008. *Assessment of the Population-Level Impacts of Potential Increases in Marine Turtle Interactions Resulting from a Hawaii Longline Association Proposal to Expand the Hawaii-Based Shallow-Set Fishery*. Pacific Islands Fisheries Science Center, National Marine Fisheries Service, National Oceanic and Atmospheric Administration, Honolulu, Hawaii.

Snover, M.L. and A.A. Hohn. 2004. Validation and interpretation of annual skeletal marks in loggerhead (*Caretta caretta*) and Kemp's ridley (*Lepidochelys kempii*) sea turtles. *Fishery Bulletin* 102:682-692.

Snover, M.L., A.A. Hohn, L.B. Crowder, and S.S. Heppell. 2007. Age and growth in Kemp's ridley sea turtles: Evidence from mark-recapture and skeletochronology. In *Biology and Conservation of Ridley Sea Turtles*, Plotkin, P.T. (ed.). Johns Hopkins University Press, Baltimore, Maryland.

Snover, M.L. and S.S. Heppell. 2009. Application of diffusion approximation for risk assessments of sea turtle populations. *Ecological Applications* 19(3):774-785.

Snover, M.L., A.A. Hohn, L.B. Crowder, and S.A. Macko. In press. Combining stable isotopes and skeletal growth marks to reconstruct ontogenetic feeding ecology and growth rate sifts in juvenile loggerhead sea turtles (*Caretta caretta*). *Endangered Species Research*.

Solis, D.S., C. Mario-Orrego, R.V. Blanco-Segura, M.R. Harfush-Melendez, E.O. Albavera-Padilla, and R.A. Valverde. 2008. Estimating arribada size: Going global. In *Proceedings of the 27th Annual Symposium on Sea Turtle Biology and Conservation*, Rees, A.F., M. Frick, A. Panagopoulou, and K. Williams (eds.). NOAA Technical Memorandum NMFS-SEFSC-569, Southeast Fisheries Science Center, National Marine Fisheries Service, National Oceanic and Atmospheric Administration, Miami, Florida.

Solow, A.R. 2001. A comment on two models of the remigration interval in marine turtles. *Journal of Theoretical Biology* 209(3):371-372.

Solow, A.R., K.A. Bjorndal, and A.B. Bolten. 2002. Annual variation in nesting numbers of marine turtles: The effect of sea surface temperature on remigration intervals. *Ecology Letters* 5(6):742-746.

Southeast Fisheries Science Center. 2010. *Sea Turtle Stranding and Salvage Network (STSSN)*. [Online]. Available: http://www.sefsc.noaa.gov/seaturtleSTSSN.jsp [2010, May 3].

Southwest Fisheries Science Center. 2008. *Green Turtle mtDNA Sequences*. [Online]. Available: http://swfsc.noaa.gov/textblock.aspx?Division=PRD&ParentMenuId=212&id=11212 [2010, May 3].

Southwood, A. and D. Kirby. 2008. *Leatherback Turtle Foraging Habitats in the North Atlantic: Implications for Fisheries Interactions*. [Online]. Available: http://www.largepelagics.unh.edu/pdf/Progress%20Reports/LPRC%20Prog%20Report%20Southwood%20Jul%2008l.pdf [2010, May 7].

Stancyk, S.E. 1982. Non-human predators of sea turtles and their control. In *Biology and Conservation of Sea Turtles*, Bjorndal, K.A. (ed.). Smithsonian Institution Press, Washington, DC.

Sullivan, P.J., J.M. Acheson, P.L. Angermeier, T. Faast, J. Flemma, C.M. Jones, E.E. Knudson, T.J. Minello, D.H. Secor, R. Wunderlich, and B.A. Zanetell. 2006. *Defining and Implementing Best Available Science for Fisheries and Environmental Science, Policy, and Management*. American Fisheries Society, Bethesda, Maryland and Estuarine Research Federation, Port Republic, Maryland.

Takahata, N. and S.R. Palumbi. 1985. Extranuclear differentiation and gene flow in the finite island model. *Genetics* 109:441-457.

Taylor, B.L., P.R. Wade, D.P. DeMaster, and J. Barlow. 2000. Incorporating uncertainty into management models for marine mammals. *Conservation Biology* 14(5):1243-1252.

Theissinger, K., N.N. FitzSimmons, C.J. Limpus, C.J. Parmenter, and A.D. Phillott. 2009. Mating system, multiple paternity and effective population size in the endemic flatback turtle (*Natator depressus*) in Australia. *Conservation Genetics* 10(2):329-346.

Thomson, D.L., S.R. Baillie, and W.J. Peach. 1997. The demography and age-specific annual survival of song thrushes during periods of population stability and decline. *Journal of Animal Ecology* 66(3):414-424.

Tiwari, M., K.A. Bjorndal, A.B. Bolten, and B.M. Bolker. 2006. Evaluation of density dependent processes and green turtle *Chelonia mydas* hatchling production at Tortuguero, Costa Rica. *Marine Ecology Progress Series* 326:283-293.

Tomás, J., P. Gozalbes, J.A. Raga, and B.J. Godley. 2008. Bycatch of loggerhead sea turtles: Insights from 14 years of strandings data. *Endangered Species Research* 5:161-169.

Troëng, S. and E. Rankin. 2005. Long-term conservation efforts contribute to positive green turtle *Chelonia mydas* nesting trend at Tortuguero, Costa Rica. *Biological Conservation* 121(1):111-116.

Troëng, S. and M. Chaloupka. 2007. Variation in adult annual survival probability and remigration intervals of sea turtles. *Marine Biology* 151(5):1721-1730.

Tucker, A.D. 2009. Eight nests recorded for a loggerhead turtle within one season. *Marine Turtle Newsletter* 124:16-17.

Tucker, A.D. 2010. Nest site fidelity and clutch frequency of loggerhead turtles are better elucidated by satellite telemetry than by nocturnal tagging efforts: Implications for stock estimation. *Journal of Experimental Marine Biology and Ecology* 383(1):48-55.

Turchin, P. 2003. *Complex Population Dynamics: A Theoretical/Empirical Synthesis*. Princeton University Press, Princeton, New Jersey.

Turtle Expert Working Group. 1998. *An Assessment of the Kemp's Ridley (*Lepidochelys kempii*) and Loggerhead (*Caretta caretta*) Sea Turtle Populations in the Western North Atlantic*. NOAA Technical Memorandum NMFS-SEFSC-409, Southeast Fisheries Science Center, National Marine Fisheries Service, National Oceanic and Atmospheric Administration, Miami, Florida.

Turtle Expert Working Group. 2000. *Assessment Update for the Kemp's Ridley and Loggerhead Sea Turtle Populations in the Western North Atlantic*. NOAA Technical Memorandum NMFS-SEFSC-444, Southeast Fisheries Science Center, National Marine Fisheries Service, National Oceanic and Atmospheric Administration, Miami, Florida.

Turtle Expert Working Group. 2007. *An Assessment of the Leatherback Turtle Population in the Atlantic Ocean*. NOAA Technical Memorandum NMFS SEFSC-555, Southeast Fisheries Science Center, National Marine Fisheries Service, National Oceanic and Atmospheric Administration, Miami, Florida.

Turtle Expert Working Group. 2009. *An Assessment of the Loggerhead Turtle Population in the Western North Atlantic Ocean*. NOAA Technical Memorandum NMFS SEFSC-575, Southeast Fisheries Science Center, National Marine Fisheries Service, National Oceanic and Atmospheric Administration, Miami, Florida.

University of Florida. 1999a. *Sea Turtle Tag Inventory*. [Online]. Available: http://accstr.ufl.edu/taginv.html [2010, May 21].

University of Florida. 1999b. *CMTTP: Cooperative Marine Turtle Tagging Program*. [Online]. Available: http://accstr.ufl.edu/cmttp.html [2010, May 21].

University of Florida. 2001. *Marine Turtle DNA Sequences*. [Online]. Available: http://accstr.ufl.edu/genetics.html [2010, May 3].

University of Florida. 2010. *ACCSTR Legacy Database Initiative: Preserving the Past to Save the Future*. [Online]. Available: http://accstr.ufl.edu/legacy.html [2010, May 21].

University of Washington. 2010. *LAMARC—Likelihood Analysis with Metropolis Algorithm using Random Coalescence*. [Online]. Available: http://evolution.genetics.washington.edu/lamarc/index.html [2010, May 12].

U.S. Department of Commerce and U.S. Department of Education. 2008. *The Shortage in the Number of Individuals with Post-Baccalaureate Degrees in Subjects Related to Fishery Science*. NOAA Technical Memorandum NMFS-F/SPO-91, Washington, DC.

Valverde, R.A. and C.E. Gates. 1999. Population surveys on mass nesting beaches. In *Research and Management Techniques for the Conservation of Sea Turtles*, Eckert, K.L., K.A. Bjorndal, F.A. Abreu-Grobois, and M. Donnelly (eds.). IUCN/SSC Marine Turtle Specialist Group Publication No. 4, International Union for the Conservation of Nature, Gland, Switzerland.

van Buskirk, J. and L.B. Crowder. 1994. Life-history variation in marine turtles. *Copeia* 1994(1):66-81.

Van Horne, B. 1983. Density as a misleading indicator of habitat quality. *Journal of Wildlife Management* 47(4):893-901.

Varis, O. and S. Kuikka. 1999. Learning Bayesian decision analysis by doing: Lessons from environmental and natural resources management. *Ecological Modelling* 119(2-3):177-195.

Vaughan, J.R. 2009. *Evaluation of Length Distributions and Growth Variance to Improve Assessment of the Loggerhead Sea Turtle (Caretta caretta)*. Master of Science Thesis, Oregon State University, Corvallis, Oregon.

Vignal, A., D. Milan, M. SanCristobal, and A. Eggen. 2002. A review on SNP and other types of molecular markers and their use in animal genetics. *Genetics Selection Evolution* 34(3):275-305.

Wade, P.R. 1998. Calculating limits to the human-caused mortality of cetaceans and pinnipeds. *Marine Mammal Science* 14(1):1-37.

Wade, P.R. and R.P. Angliss. 1997. *Guidelines for Assessing Marine Mammal Stocks: Report of the GAMMS Workshop April 3-5, 1996, Seattle, Washington.* NOAA Technical Memorandum NMFS-OPR-12, Northwest Fisheries Science Center, National Marine Fisheries Service, National Oceanic and Atmospheric Administration, Seattle, Washington.

Wallace, B.P., S.S. Heppell, R.L. Lewison, S. Kelez, and L.B. Crowder. 2008. Impacts of fisheries bycatch on loggerhead turtles worldwide inferred from reproductive value analyses. *Journal of Applied Ecology* 45(4):1076-1085.

Wallace, B.P., E.M. Finkbeiner, A.D. DiMatteo, and S. Helmbrect. 2009a. A framework for global threats evaluation, gap analyses, diversity hotspots assessment, and conservation priority setting for marine turtles. In *Proceedings of the First International Marine Conservation Congress, Society of Conservation Biology*, Washington, DC.

Wallace, B., E.M. Finkbeiner, S. Helmbrecht, and A.D. DiaMatteo. 2009b. *Developing Regional Management Units for Marine Turtles Worldwide.* Presentation at the International Marine Conservation Congress, Washington, DC.

Walters, C.J., V. Christensen, and D. Pauly. 1997. Structuring dynamic models of exploited ecosystems from trophic mass-balance assessments. *Reviews in Fish Biology and Fisheries* 7(2):139-172.

Waples, R.S. 1991. *Definition of "Species" under the Endangered Species Act: Application to Pacific Salmon.* NOAA Technical Memorandum F/NWC-194, Northwest Fisheries Science Center, National Marine Fisheries Service, National Oceanic and Atmospheric Administration, Seattle, Washington.

Waples, R.S. 1995. Evolutionarily significant units and the conservation of biological diversity under the Endangered Species Act. In *Evolution and the Aquatic Ecosystem: Defining Unique Units in Population Conservation*, Nielsen, J.L. (ed.). American Fisheries Society Symposium 17, American Fisheries Society, Bethesda, Maryland.

Weishampel, J.F., D.A. Bagley, and L.M. Ehrhart. 2004. Earlier nesting by loggerhead sea turtles following sea surface warming. *Global Change Biology* 10:1424-1427.

Wetherall, J., G.H. Balazs, R. Tokunaga, and M. Yong. 1993. Bycatch of marine turtles in North Pacific high-seas driftnet fisheries and impacts on the stocks. *North Pacific Fisheries Commission Bulletin* 53(3):519-538.

Whelan, C. and J. Wyneken. 2007. Estimating predation levels and site-specific survival of hatchling loggerhead sea turtles (*Caretta caretta*) from south Florida beaches. *Copeia* 2007(3):745-754.

Whitlock, M.C., M.A. McPeek, M.D. Rausher, L. Rieseberg, and A.J. Moore. 2010. Data archiving. *The American Naturalist* 175(2):145-146.

Whiting, A.U., M. Chaloupka, and C.J. Limpus. 2008. Sampling error for hatchling turtle measurements: Probing a rule-of-thumb. *Copeia* 2008(4):889-896.

Wibbels, T. 2003. Critical approaches to sex determination in sea turtles. In *The Biology of Sea Turtles, Volume II*, Lutz, P.L., J.A. Musick, and J. Wyneken (eds.). CRC Press, Boca Raton, Florida.

Wiens, J.A. 1989. Spatial scaling in ecology. *Functional Ecology* 3(4):385-397.

Wilson, G.A. and B. Rannala. 2003. Bayesian inference of recent migration rates using multilocus genotypes. *Genetics* 163(3):1177-1191.

Wilson, R.P., E.L.C. Shepard, and N. Liebsch. 2008. Prying into the intimate details of animal lives: Use of a daily diary on animals. *Endangered Species Research* 4(1-2):123-137.

Witherington, B.E. 1991. Orientation of hatchling loggerhead turtles at sea off artificially lighted and dark beaches. *Journal of Experimental Marine Biology and Ecology* 149(1):1-11.

Witherington, B.E. 2002. Ecology of neonate loggerhead turtles inhabiting lines of downwelling near a Gulf Stream front. *Marine Biology* 140(4):843-853.

Witherington, B.E. and L. Ehrhart. 1989. Hypothermic stunning and mortality of marine turtles in the Indian River Lagoon System, Florida. *Copeia* 1989(3):696-703.

Witherington, B.E., P. Kubilis, B. Brost, and A.B. Meylan. 2009. Decreasing annual nest counts in a globally important loggerhead sea turtle population. *Ecological Applications* 19(1):30-54.

Witzell, W.N. 1994. The origin, evolution and demise of the U.S. sea turtle fisheries. *Marine Fisheries Review* 56(4):8-23.

Wood, F.E. and J.R. Wood. 1985. Release of "living tag" marked green turtles in the Cayman Islands. *Marine Turtle Newsletter* 32:5-6.

Wood, J.R. and F.E. Wood. 1981. Growth and digestibility for the green turtle (*Chelonia mydas*) fed diets containing varying protein levels. *Aquaculture* 25(2-3):269-274.

Wright, S. 1943. Isolation by distance. *Genetics* 28(2):114-138.

Wyneken, J., S.P. Epperly, L.B. Crowder, J. Vaughan, and K.B. Esper. 2007. Determining sex in posthatchling loggerhead sea turtles using multiple gonadal and accessory duct characteristics. *Herpetologica* 63(1):19-30.

Xu, S., C.J. Kobak, and P.E. Smouse. 1994. Constrained least squares estimation of mixed population stock composition from mtDNA haplotype frequency data. *Canadian Journal of Fisheries and Aquatic Science* 51(2):417-425.

Yalçın-Özdilek, S. and B. Sönmez. 2006. Some properties of new nesting areas of sea turtles in north-eastern Mediterranean situated on the extension of the Samanda Beach, Turkey. *Journal of Environmental Biology* 27(3):537-544.

Zug, G.R., A.H. Wynn, and C.A. Ruckdeschel. 1986. Age determination of loggerhead sea turtles, *Caretta caretta*, by incremental growth marks in the skeleton. *Smithsonian Contributions to Zoology* 427:1-34.

Zug, G.R., G.H. Balazs, and J. Wetherall. 1995. Growth in juvenile loggerhead sea turtles (*Caretta caretta*) in the North Pacific pelagic habitat. *Copeia* 1995(2):484-487.

Appendixes

A

Brief History of Alternative Genetic Markers

The first wave of molecular genetic data on sea turtles included a variety of techniques, during a period when DNA sequence data were still expensive and laborious to obtain. For the purposes of sea turtle population studies, those techniques have largely been replaced. However, it is notable that the conclusions based on them have been confirmed (for the most part) with newer technologies based on the polymerase chain reaction (PCR), a genetic-analysis technique that is used to amplify pieces of DNA and generates millions of copies of a particular sequence. As explained in Chapter 2, mitochondrial DNA control-region sequences and hypervariable microsatellites[1] are the genetic-analysis techniques of choice for sea-turtle population assessment and are likely to remain the primary ones for the next decade (Bowen and Karl, 2007). One promising genetic-analysis technique that has not yet been applied to sea turtles is single-nucleotide polymorphisms (a DNA-sequence variation that can occur among members of the same species; Vignal et al., 2002), which require extensive nuclear DNA sequence information to identify variable sites throughout the genome (Lee, 2008).

Restriction Fragment Length Polymorphisms (Bowen et al., 1992)—This technique takes advantage of a suite of restriction enzymes,[2] which can

[1] Also known as DNA fingerprints, these are highly variable DNA sequences that occur in short repeats, such as GAGAGAGAGA. The number of repeats can vary from a few to over 30 so it is possible to have many variants at this hypervariable region.

[2] These are useful for cutting genomes into fragments that are small enough to manipulate in cloning or DNA sequencing.

cut DNA at specific four–, five–, or six–base-pair sequences. For example, the enzyme *EcoR1* (a restriction enzyme derived from the bacterium *Escherichia coli*) cuts DNA at sites that contain the nucleotide sequence GAATTC. This is a quick and inexpensive way to get sequence information and was widely used in population-genetics studies before the advent of PCR-based sequencing technology. The technique is highly repeatable and robust but has largely been replaced by direct DNA sequencing.

Anonymous Single-Copy Nuclear DNA (Karl et al., 1992)—This technique requires cloning and sequencing fragments of DNA. On the basis of the clones, variation in the nuclear genome can be resolved and characterized. The requirement of cloning (like microsatellites; see Chapter 2) makes this an expensive and labor-intensive approach to initialize, but it is robust and repeatable (Karl and Avise, 1993). In population genetic studies, it is largely replaced by microsatellite methods but has broad applications in phylogeography and phylogenetic studies.[3]

Minisatellites (Peare and Parker, 1996)—These are the first generation of "DNA fingerprints" and consist of short repeat sequences[4] of about 10–60 base pairs that occur in variable copy number, in hundreds of locations in the genome. They are detected with a fluorescent or radioactive probe and can be variable enough to distinguish individuals (Jeffreys, 2005). However, they can be difficult to interpret and have largely been replaced by microsatellites in population-genetics studies.

Random Amplification of Polymorphic DNA (Schroth et al., 1996)—This technique uses PCR primers to amplify short segments of the genome randomly; they are then separated and visualized with gel electrophoresis.[5] It has the advantage of not requiring prior knowledge of the genome (sequence data) to design primers. However, it is not widely used in population-genetics studies because of problems with interpretation and repeatability.

---

[3] Phylogeography focuses on the geographic distribution of genetic variation, usually at the level of species and genera. Phylogeographic studies often reveal molecular evolutionary separations below the species level, as is the case with green turtles (*Chelonia mydas*; Bowen et al., 1992). Phylogenetics is the study of evolutionary history, usually describing relationships among species, genera, and higher taxonomic categories in the format of trees.

[4] These are DNA segments that repeat the same sequence multiple times. They are prone to duplication during cell replication and therefore can produce highly variable genetic markers.

[5] This is a method of separating DNA fragments by size. The DNA or protein is inserted into a gelatin slab, and an electric current is run through the gelatin to move fragments toward either the positive or negative end.

B

Population-Structure Models

ISLAND MODEL

The island model is a basic model of division of a species into isolated "islands". The organisms breed randomly within an island, and migrants are drawn randomly from each island. Because of the division and isolation, the heterozygosity (when two different alleles occupy the gene's positions, or loci, on a pair of chromosomes) of the entire group is lower than would be expected if there were random mating among all members of the species (Wright, 1943).

ISOLATION BY DISTANCE

When a species inhabits a large geographic area, genotype frequencies may change gradually in space in a way that is not due to physical barriers. Such isolation occurs because the geographic area is much greater than an individual's migration distance. This model reflects the loss of heterogeneity that results when individuals breed with their neighbors and produce genetic differentiation across the range. At the ends of the range, there is smaller genetics correlation than in nearby localities (Wright, 1943). Dethmers et al. (2006) observed isolation by distance—greater than 2,000 km—in west Pacific green turtles (*Chelonia mydas*).

STEPPING-STONE

This model "assumes that the entire population is subdivided into colonies and the migration of individuals in each generation is restricted to nearby colonies" (Kimura and Weiss, 1964). Thus it is a special case of isolation by distance.

METAPOPULATION

Sewall Wright and others developed the ideas of spatial structure in populations relative to their genetics. Levins (1969, 1970) reframed the effects of spatial structure to the population dynamics and ecology of a species. His model concentrated on the effects of extinction and recolonization of local populations on the persistence of a species. Hanski and Simberloff (1997) have developed the ideas further since the 1980s to emphasize the effect of migration and connectivity on the vital rates of local populations and how spatial heterogeneity can act to protect a species from extinction. Modern theory does not necessarily assume that local populations will go extinct and allows that there can be substantial migration between them. However, the ramifications of habitat fragmentation on formation of metapopulations has not been fully developed (Jones, 2006).

C

Committee and Staff Biographies

COMMITTEE

Karen A. Bjorndal (*Chair*) is a professor of biology and the director of the Archie Carr Center for Sea Turtle Research at the University of Florida in Gainesville. She received a Ph.D. in zoology from the University of Florida. Dr. Bjorndal served as the chair of the Marine Turtle Specialist Group of the International Union for the Conservation of Nature for 12 years. She is a member of the Scientific Advisory Council of the Bahamas National Trust and served as president of the Comparative Nutrition Society. Her research includes sea turtle demographics, feeding ecology, growth rates, and nutrition. Dr. Bjorndal was a member of the National Research Council Committee on Sea Turtle Conservation, which issued *Decline of the Sea Turtles: Causes and Prevention* in 1990.

Brian W. Bowen is an associate researcher at the Hawaii Institute of Marine Biology, part of the University of Hawaii at Manoa. He received a Ph.D. in genetics from the University of Georgia in 1992. Dr. Bowen's research focuses on the phylogeography and conservation genetics of marine vertebrates. His research program is designed to serve conservation goals by illuminating the evolutionary processes that generate biodiversity. Dr. Bowen is a member of the American Association for the Advancement of Science, the American Genetic Association, the Society for Conservation Biology, and the Society for the Study of Evolution. He is an author of the textbook *Diversity of Fishes, Second Edition*. He has held many editorial positions and is now an associate editor of the *Journal* *of Heredity* and a member of the Editorial Review Board of *Molecular Ecology*.

Milani Chaloupka runs Ecological Modeling Services Pty Ltd., an international research company that provides statistical and mathematical consulting on ecological and economic issues for a wide array of groups, including industry, government, academe, and nongovernmental organizations. Dr. Chaloupka has a Ph.D. in marine ecology from the University of Queensland in Australia. His expertise is in statistical and mathematical modeling of complex ecological systems, including the development of interactive stochastic computer simulations of endangered-species population dynamics. He is the chair of the Sea Turtle Advisory Committee of the Western Pacific Regional Fishery Management Council, chair of the Marine Turtle Red List Authority, and vice-chair of the Marine Turtle Specialist Group.

Larry B. Crowder is a professor of marine ecology at the Nicholas School for the Environment of Duke University. He completed his doctoral studies in zoology at Michigan State University. Dr. Crowder's research centers on predation and food-web interactions, mechanisms underlying recruitment variation in fish, and population modeling in conservation biology. Dr. Crowder is currently engaged in more extensive programs in marine conservation, including endangered species and fishery conflicts, especially bycatch in fishing gear. Dr. Crowder is a former member of the Ocean Studies Board and has served on several National Research Council committees.

Selina S. Heppell is an associate professor in the Department of Fisheries and Wildlife at Oregon State University. She earned a Ph.D. in zoology from Duke University. Dr. Heppell's research focuses on sea turtles, sharks, sturgeon, and U.S. west coast rockfish, primarily using computer models and simulations to understand how populations respond to human activities and to guide research and management policy toward their recovery. She was an Aldo Leopold Leadership Program Fellow in 2006.

Cynthia M. Jones is a professor, an eminent scholar, and the director of the Center for Quantitative Fisheries Ecology at Old Dominion University. She received a Ph.D. in oceanography from the University of Rhode Island. Dr. Jones is a recognized expert in fishery ecology and population dynamics, and her recent research has explored such topics as elemental analysis of adult and juvenile fish to investigate natal homing and connectivity in a marine fish metapopulation. She has served on the Ocean Studies Board and several National Research Council committees.

Molly E. Lutcavage is a research professor and the director of the Large Pelagics Research Center in the Department of Natural Resources and the Environment of the University of Massachusetts Amherst. Dr. Lutcavage received her M.S. from Virginia Institute of Marine Science at the College of William and Mary and her Ph.D. in biological oceanography from the University of Miami. Her research emphasizes population biology, physiological ecology, and conservation of large pelagic species, particularly tunas, billfish and sea turtles. With her colleagues, Dr. Lutcavage helped to develop methods of electronic tagging and tracking of large marine animals. Her current interests include development of fishery-independent detection and assessment methods for large pelagic species and cooperative research approaches for fisheries.

Andrew R. Solow is a senior scientist and the director of the Marine Policy Center of the Woods Hole Oceanographic Institution. He received his Ph.D. from Stanford University. His research is in environmental and ecological statistics. Dr. Solow has served on several National Research Council committees.

Blair E. Witherington is a research scientist with the Florida Fish and Wildlife Research Institute, where he has worked since 1992 on sea-turtle biology and conservation. Dr. Witherington received a Ph.D. from the University of Florida. He is an adjunct assistant professor in the Department of Zoology of the University of Florida, serves as a Fulbright senior specialist in biology, has served as president of the 20th International Sea Turtle Symposium, and is a member of the Marine Turtle Specialist Group of the International Union for Conservation of Nature as vice-chair of the Northwest Atlantic region.

STAFF

Jodi Bostrom is an associate program officer with the Ocean Studies Board. She earned a B.S. in zoology from the University of Wisconsin-Madison and an M.S. in environmental science from American University. Since starting with the Ocean Studies Board in May 1999, Ms. Bostrom has worked on several studies pertaining to coastal restoration, fishery policy, marine mammals and noise, nutrient over-enrichment, ocean exploration, capacity building for oceans and coasts, land-based marine debris, and best practices for shellfish aquaculture.

Susan Park (until December 2009) was a senior program officer with the Ocean Studies Board until the end of 2009. She received her Ph.D. in oceanography from the University of Delaware in 2004. Dr. Park was a Christine Mirzayan Science and Technology Graduate Policy Fellow with the Ocean Studies Board in 2002 and joined the staff in 2006. She worked on several reports, including *Nonnative Oysters in the Chesapeake Bay*, *Review of Recreational Fisheries Survey Methods*, *Dynamic Changes in Marine Ecosystems*, *A Review of the Ocean Research Priorities Plan and Implementation Strategy*, and *Tackling Marine Debris in the 21st Century*. Before joining the Ocean Studies Board, she worked on aquatic invasive-species management with the Massachusetts Office of Coastal Zone Management and the Northeast Aquatic Nuisance Species Panel. She is now the assistant director for research at Virginia Sea Grant.

David Policansky has a B.A. in biology from Stanford University and an M.S. and a Ph.D. in biology from the University of Oregon. He has taught introductory biology, genetics, ichthyology, evolution, ecology, and graduate seminars. He is a scholar with the Board on Environmental Studies and Toxicology, where he directs studies on applied ecology and natural-resource management. He is a member of the Ecological Society of America and the American Fisheries Society and chairs the advisory council for the University of Alaska's School of Fisheries. He was a 2001 Harriman Scholar on the retracing of the 1899 Harriman Alaska Expedition. His interests include genetics; evolution; ecology, including the effects of fishing on fish populations; ecological risk assessment; natural-resource management; and how science is used in informing policy. He has directed more than 30 projects at the National Research Council on natural resources and ecological risk assessment, including reports on the Endangered Species Act; on salmon in the Pacific Northwest, Maine, and Alaska; on wetlands delineation; on enhancing water supplies in the Middle East; on cumulative environmental effects of oil and gas activities on Alaska's North Slope; on ecological indicators; on environmental effects of wind-energy projects; and on ecosystem-based approaches to the management of marine fisheries. He has published about 35 papers, book chapters, and book reviews, most recently on fisheries, the role of science in decision making, and common-property resources.

Jeremy Justice is a senior program assistant with the Ocean Studies Board. He earned a B.A. in international and area studies from the University of Oklahoma in 2008. Since joining the staff in October 2008, Mr. Justice has worked on *Science at Sea: Meeting Future Oceanographic Goals with a Robust Academic Research Fleet* and *Ecosystem Concepts for Sustainable Bivalve Mariculture*, in addition to this report.